Drafting and Design for Architecture

Solutions Manual

Dana J. Hepler
Principal: Hepler Associates PC
New York City and Massapequa Park, NY
Adjunct Professor of Architecture
New York Institute of Technology

Paul R. Wallach
Architectural Instructor—Cañada College
Technical Writer and Consultant
Burlingame, CA

Donald E. Hepler
President: Technical Writing and Design Service Inc.
Somers, CT

Africa • Australia • Canada • Denmark • Japan • Mexico • New Zealand • Philippines • Puerto Rico • Singapore • Spain • United Kingdom • United States

Drafting and Design for Architecture Solutions Manual

Dana J. Hepler
Paul R. Wallach
Donald E. Hepler

Vice President, Technology and Trades SBU:
Alar Elken

Editorial Director:
Sandy Clark

Senior Acquisitions Editor:
James DeVoe

Senior Development Editor:
John Fisher

Marketing Director:
Dave Garza

Channel Manager:
Bill Lawrenson

Marketing Coordinator:
Stacey Wiktorek

Production Director:
Mary Ellen Black

Production Manager:
Andrew Crouth

Production Editor:
Stacy Masucci

Art/Design Specialist:
Mary Beth Vought

Technology Project Manager:
Kevin Smith

Technology Project Specialist:
Linda Verde

Editorial Assistant:
Tom Best

COPYRIGHT © 2006 Thomson Delmar Learning. Thomson, the Star Logo, and Delmar Learning are trademarks used herein under license.

Printed in Canada
1 2 3 4 5 XX 08 07 06 05

For more information contact
Thomson Delmar Learning
Executive Woods
5 Maxwell Drive, PO Box 8007,
Clifton Park, NY 12065-8007
Or find us on the World Wide Web at
www.delmarlearning.com

ALL RIGHTS RESERVED. No part of this work covered by the copyright hereon may be reproduced in any form or by any means—graphic, electronic, or mechanical, including photocopying, recording, taping, Web distribution, or information storage and retrieval systems—without the written permission of the publisher.

For permission to use material from the text or product, contact us by
Tel. (800) 730-2214
Fax (800) 730-2215
www.thomsonrights.com

Library of Congress Cataloging-in-Publication Data:
Card Number:
ISBN: 1-4018-7998-5

Notice to the Reader

Publisher does not warrant or guarantee any of the products described herein or perform any independent analysis in connection with any of the product information contained herein. Publisher does not assume, and expressly disclaims, any obligation to obtain and include information other than that provided to it by the manufacturer.

The reader is expressly warned to consider and adopt all safety precautions that might be indicated by the activities herein and to avoid all potential hazards. By following the instructions contained herein, the reader willingly assumes all risks in connection with such instructions.

The publisher makes no representation or warranties of any kind, including but not limited to the warranties of fitness for particular purpose or merchantability, nor are any such representations implied with respect to the material set forth herein, and the publisher takes no responsibility with respect to such material. The publisher shall not be liable for any special, consequential, or exemplary damages resulting, in whole or part, from the readers' use of, or reliance upon, this material.

CONTENTS

INTRODUCTION

This Instructor's Guide is prepared to accompany the text *Drafting and Design for Architecture* and the related *Student Workbook*. Contents cover suggestions for developing and managing courses in Architectural Drafting and Design including class organization, lesson plans, course outlines, test banks, student skill development, and teaching strategies. Answers to text chapter reviews and student workbook exercises are included. Power point illustrations, handout materials, and a full size set of architectural drawings are provided to reinforce and augment text coverage. Topics not appropriate or too voluminous for the student text or workbook, such as assessing student performance, competency-based criteria, Feng Shui, metric applications, and professional references are also presented. This Instructor's Guide is available on a CD-ROM titled: *The Instructor's e-resource CD.*

DRAFTING and DESIGN for ARCHITECTURE is a basic architectural text designed for a beginning course in architectural drafting and design. This text is intended to help students learn the fundamental skills and concepts necessary for architectural planning, designing, and drawing. The main thrust of this text is to show students how to prepare architectural drawings that will effectively and accurately communicate ideas and designs to clients, contractors, and construction personnel.

Information in this text is based on the competencies necessary to understand architectural planning and designing and to illustrate ideas through drafting. The instructional material is presented to expose the student to concepts by beginning gradually with simple examples and progressing to more complete applications. Exposure in the text is from the simple to the complex and from the familiar to the unfamiliar. Each topic is first presented with only as much detail as required for student comprehension. Each time the topic is reintroduced, more technical depth is added, until the maximum complexity is reached. In this manner, symbols, terms, and concepts are defined where first introduced and also repeated wherever the opportunity exists. Through this approach, students receive reinforcement of what they have learned in previous units while continuing to expand their knowledge.

The text is designed to provide a foundation for the study of many related fields in architecture, construction, and engineering. While covering the entire text is helpful in preparing for many careers, some topics are more relevant than others to some careers. The following chart shows the text chapters that apply most significantly to major architecture, construction, and engineering careers.

TEXT CHAPTERS RECOMMENDED FOR CAREER PREPARATION

	Architectural Drafter	Architect	Landscape Architect	Civil Engineer	Structural Engineer	Architectural Designer	Contractor	Carpenter	Mason	Plumber	Electrician	HVAC Contractor	Interior Designer	Home Owner
Part 1: Introduction to Architecture														
1 - Architectural History and Styles		x	x	x		x	x						x	x
2 - Fundamentals of Design	x	x	x	x	x	x	x						x	x
Part 2: Architectural Drafting Fundamentals														
3 - Drafting Scales and Instruments	x	x	x	x	x	x	x						x	
4 - Architectural Drafting Conventions	x	x	x	x	x	x	x						x	
5 - Introduction to Computer-Aided Drafting and Design	x	x	x	x	x	x	x						x	
Part 3: Basic Area Design														
6 - Environmental Design Factors		x	x	x		x	x					x	x	x
7 - Indoor Living Areas		x	x			x							x	x
8 - Outdoor Living Areas		x	x			x							x	x
9 - Traffic Areas and Patterns		x	x			x							x	x
10 - Kitchens		x				x							x	x
11 - General Service Areas		x				x							x	x
12 - Sleeping Areas		x				x							x	x
Part 4: Basic Architectural Drawings														
13 - Designing Floor Plans		x			x	x							x	
14 - Drawing Floor Plans	x	x			x	x	x						x	
15 - Designing Elevations		x				x								
16 - Drawing Elevations	x	x				x	x							
17 - Sectional, Detail, and Cabinetry Drawings	x	x				x	x						x	
18 - Site Development Plans	x	x	x	x		x	x							
Part 5: Presentation Methods														
19 - Pictorial Drawings	x	x	x			x							x	
20 - Architectural Renderings	x	x	x			x							x	
21 - Architectural Models	x	x	x	x	x	x	x	x					x	

	Architectural Drafter	Architect	Landscape Architect	Civil Engineer	Structural Engineer	Architectural Designer	Contractor	Carpenter	Mason	Plumber	Electrician	HAVC Contractor	Interior Designer	Home Owner
Part 6: Foundations and Construction Systems														
22 - Principles of Construction	x	x	x	x	x	x	x	x	x	x		x	x	
23 - Foundations and Fireplace Structures	x	x	x	x	x	x	x	x	x					
24 - Wood-Frame Systems	x	x	x	x		x	x	x						
25 - Masonry and Concrete Systems	x	x	x	x		x	x		x					
26 - Steel and Reinforced-Concrete Systems	x	x	x	x		x	x	X	x					
27 - Disaster Prevention Design	x	x		x		x	x	x	x	x	x	x		
Part 7: Framing Systems														
28 - Floor Framing Drawings	x	x		x	x	x		x		x	x	x		
29 - Wall Framing Drawings	x	x		x	x	x		x		x	x	x		
30 - Roof Framing Drawings	x	x		x	x	x		x			x	x		
Part 8: Electrical and Mechanical Design and Drawings														
31 - Electrical Design and Drawings	x	x	x		x	x	x				x	x	x	
32 - Comfort-Control Systems (HVAC)	x	x			x	x	x				x	x		
33 - Plumbing Drawings	x	x		x	x	x	x			x		x		
Part 9: Checking Plans and Using Support Services														
34 - Drawing Coordination and Checking	x	x	x	x	x	x	x	x	x	x	x	x	x	
35 - Schedules and Specifications	x	x	x	x	x	x	x	x	x	x	x	x	x	
36 - Building Costs and Financial Planning		x	x	x	x	x	x	x	x	x	x	x	x	x
37 - Codes and Legal Documents		x	x	x	x	x	x	x	x	x	x	x	x	x

OVERVIEW

The *Drafting and Design for Architecture* program includes three components for teaching and learning:

- Student Textbook—a text that covers the basic principles of drafting and design for architecture.
- Instructor's Guide—an extensive collection of resource and reference materials and other helpful information.
- Student Workbook—drawing exercises that help students apply the knowledge and techniques gained in class.
- E.Resource—Each of these components is described in more detail in this section.

The Student Text

The *Drafting and Design for Architecture* textbook has been revised and expanded to include the latest technological information, methodology, and standards relating to architectural drafting, design, and construction. The chapters on computer-aided drafting (CAD), CAD applications, site development, drafting conventions, ergonomics, design factors and procedures, construction systems, related careers, and schedules have been completely revised. All new and expanded information adheres to contemporary architectural practices.

Philosophy

Advancing technology and the dynamics of change are creating new challenges for students entering the workplace. Meeting these challenges requires continual retraining and upgrading of skills. Students must understand the essential technical concepts and scientific principles upon which their work is based and their careers are built. Contemporary technology programs should provide the opportunity for students to develop:

- an awareness of the importance of human relations.
- an insight into technology and its place in our society.
- problem-solving abilities related to human and technological processes.

The study of architectural drafting technologies should be integrated with the development of competencies in mathematics, science, language arts, history, and social science. Students should learn through both individual and group activities in which they use all the relevant equipment, techniques, materials, and processes.

Purpose

The *Drafting and Design for Architecture* program is intended for use in a first course in architectural design or construction drafting. It will help students learn the fundamental skills and concepts necessary for planning, designing, and drawing. Its main thrust is to show students how to prepare architectural drawings that will effectively and accurately communicate ideas and designs to clients, contractors, and construction personnel.

Because a course in basic drafting normally precedes a course in architectural drafting, only the principles and design practices as applied to architectural drawings are presented in the text. Orthographic principles are reviewed when appropriate.

Organization

The chapters in the *Drafting and Design for Architecture* student text are set up to be used consecutively; however, following a different sequence may be more suitable for certain classes. For instance, if the emphasis is to be placed on developing basic architectural drafting skills while de-emphasizing design, Parts 1 and 3 may be omitted. Or, if classes will be oriented toward the construction phase of architecture, Part 4 or Part 6 may be a more logical place to begin.

Because drawing and reading construction documents are skills vital to many architectural and related fields, this text also can serve as an introduction to construction. Each part begins with a profile of a professional in architecture or construction. Also, part ten deals exclusively with careers in both fields.

Instructor's Guide

The Instructor's Guide is designed to provide helpful resources and instructional aids that make teaching easier and more efficient.

The emphasis in the text is on the preparation of drawings that accurately and effectively communicate ideas and designs to clients. The Instructor's Guide provides tools that the instructor needs to help students understand these concepts and develop the drawing skills required. (ISBN 1-4018-7998-5) These tools include:

- **Teaching Strategies.** This section is organized by textbook chapter and contains information, answers to exercises, and teaching suggestions related to the material in those chapters.

- **Using the Student Workbook.** This section gives suggestions for using the drawing exercises and CAD tutorials in the workbook to help students improve their drafting skills while applying principles of architectural design.
- **Interpreting Architectural Drawings Test.** This section includes tests designed to determine a student's understanding of the relationship among drawings in a set. An answer key is also included.
- **Architectural Drawing Set.** A full-size (1/4″ = 1′-0″) set of residential plans is provided to demonstrate the actual size, scope, and relationship among drawings in a set complete with floor plans, elevations, sections, and details.

The Instructor's E.Resource CD

This educational resource creates a truly electronic classroom. It is a CD-ROM containing tools and instructional resources that enrich the classroom and make the instructor's preparation time shorter. The elements of the e.resource link directly to the text to provide a unified instructional system. With the e.resource you can spend your time teaching, not preparing to teach. ISBN 1-4018-7996-9.

Features contained in the e.resource include the following:

- **Course Outlines.** The skills and material to be learned are outlined by text chapters for a nine-, eighteen-, and thirty-six-week course. Recommended electives for vocational and professional career preparation are also provided.
- **Program Development.** This section includes instructional suggestions as well as information on classroom organization and developing student skills. Work–school relationships, methods of conducting a course, and out-of-class activities are also covered.
- **Handout Masters.** The handout masters are designed to be printed, photocopied, and distributed to students. They extend and enrich the text content—sometimes with a drawing, sometimes with written information. Each handout is keyed to a specific text chapter. In the "Teaching Strategies" section, you will find tips for using the handouts in your classes.
- **PowerPoint Presentations.** These slides provide the basis for a lecture outline to present concepts and material. Key points and concepts can be graphically highlighted for student retention.
- **Optical Image Library.** This database of key images (all in full color) taken from the text can be used in lecture presentations, tests and quizzes, and PowerPoint presentations. Additional Image Masters and Color Image Masters, which tie directly to lesson plans provided in the "Teaching Strategies" section, are also provided.
- **Exam View Test Bank.** Questions of varying levels of difficulty are provided in true/false, multiple choice, fill-in-the-blank, and short answer formats so you can assess student comprehension. This versatile tool enables the instructor to manipulate the data to create original tests.
- **Feng Shui.** Because of the increased popularity of this approach to architectural design, an introduction is provided that includes a collection of feng shui design features.
- **Metric Applications in Architecture.** An introduction to the metric system as applied to architectural drafting plus metric drafting conventions is presented here. A set of metric working drawings and related questions and answers is also provided.
- **Professional References.** These references include professional organizations related to all phases of architecture, construction, and engineering. A list of accredited schools of architecture in the United States is provided.

The Student Workbook

The contents of the Student Workbook are related to the text by chapter and include 250 drawing exercises to be used as tests, assignments, or teaching aids. Two AutoCAD tutorials (AutoCAD 2004) to be used by students to complete two basic plans are also provided. ISBN 1-4018-7997-7.

TEACHING STRATEGIES

This section contains a lesson plan for each chapter in the text. This includes an outline of the lesson, objectives, suggestions for image master use, and evaluation checks. Answers to end-of-chapter exercises are also provided.

Date	M	Tu	W	Th	F

Lesson Plan for Chapter 1 (Text pages 1–16)

Architectural History and Styles

ESSENTIAL ELEMENTS

For a complete listing of the essential elements, see the "Program Development" section in this guide. Essential elements for the course: 2.1, 2.2, 2.3, 5.3, 6.10, 6.12.

CHAPTER 1 OUTLINE

The Development of Architectural Forms
- Post and Lintel
- The Arch
- The Vault
- The Dome
- The Gothic Arch

How Architectural Styles Develop

Influences on Early American Architecture
- English Architecture
- French Architecture
- Spanish and Italian Architecture

Early American Styles
- New England Colonial
- Mid-Atlantic Colonial
- Dutch Colonial
- Southern Colonial

Later American Styles
- Victorian
- Ranch Style
- Influences on Contemporary Styles

New Challenges in Architecture

OBJECTIVES

Following are the objectives that appear at the beginning of Chapter 1 in the textbook. Students are expected to learn to:

- recognize historical architectural styles and identify several distinct characteristics of each style.
- relate how the development of materials and construction methods influenced architectural styles.

USING THE COLOR IMAGE MASTERS AND IMAGE MASTERS

The following color image masters (CIM), image masters (IM), and handout masters (HO) are appropriate for use with this chapter. Teaching suggestions are included.

CIM-1A: New England Colonial Home

- Show this transparency as typical of New England colonial design. Note the symmetrical balance created by the double-hung windows and shutters, main entrance in the center, and horizontal siding.
- Have students point out other features common to New England colonial designs.

CIM-1B: Southern Colonial Home

- Show as an example of Southern colonial architecture, featuring two-story columns and bilevel verandas. Note the formal balance of this design.
- Bring pictures of ante-bellum architecture to class. Ask students to compare those designs to this modern one. What features have been changed?

IM-1A Historic Columns

- Use to show the difference between ancient column design in more detail.

IM-1B Historic Structures

- This illustration covers a wide span of building history.

EVALUATION

1. Assign the exercises at the end of the textbook chapter. When applicable, check students' answers against the answers given in the next section.
2. Test student mastery of factual material by using Chapter 1 Review found in the "Chapter Reviews" section of this guide.
3. In a check for competence, students should demonstrate understanding of:
   - the history of architecture and the development and use of the bearing wall, post-and-lintel, arch and keystone, vault, and dome.
   - the significance of technological advances in materials and construction methods.
4. In a check for competence, students should be able to perform these skills:
   - make rough sketches of a post-and-lintel, arch, barrel vault, dome, and Gothic arch.
   - list the new developments in technology that have had the greatest influence on architecture and building.

ANSWERS TO END-OF-CHAPTER EXERCISES

Many end-of-chapter exercises ask students to perform an activity, such as draw a floor plan or photograph local structures. As a result, no "answer" is required and none is given here.

1. Answers will vary; however, some reference should be made to a style as evolving from a particular culture's solutions to building problems.
2. A beam across two posts. Examples and comparisons will vary but may include Greek and Roman columns and Oriental designs. Arches consist of many smaller, lighter blocks supported by a keystone in the center.
3. No answer required.
4. England and France. English styles feature high-pitched roofs, massive chimneys, half-timber siding, small windows, and exterior stone walls. French styles feature Mansard and steeply pitched hip roofs, long projecting windows, corner quoins, curved lintels, and towers.
5. Descriptions will vary. Contemporary designs feature new materials such as plastic or glass; new construction methods; fewer design restrictions; simpler, bolder, less cluttered designs; use of various geometric shapes.
6. Answers will vary.
7. Answers will vary.
8. Answers will vary.

Date	M	Tu	W	Th	F

Lesson Plan for Chapter 2 (Text pages 17–26)

Fundamentals of Design

ESSENTIAL ELEMENTS

For a complete listing of the essential elements, see the "Program Development" section in this guide. Essential elements for the course: 5.3, 6.9, 6.10

CHAPTER 2 OUTLINE

Architecture and Design
- Form Follows Function
- Interior Design
- Creativity in Architectural Design

Elements of Design
- Line
- Form
- Space
- Color
- Light and Shadow
- Texture and Materials

Principles of Design
- Balance
- Rhythm and Repetition
- Emphasis and Subordination
- Proportion
- Unity
- Variety and Opposition
- Transition

OBJECTIVES

Following are the objectives that appear at the beginning of Chapter 2 in the textbook. Students are expected to learn to:

- relate design concepts to architecture.
- identify six elements of design.
- apply design principles to a work of architecture.

USING THE COLOR IMAGE MASTERS AND IMAGE MASTERS

The following color image masters (CIM), image masters (IM), and handout masters (HO) are appropriate for use with this chapter. Teaching suggestions are included.

CIM-2A: Stone Residence

- Have students find examples of the principles and elements of design in this structure.
- Ask students what effect would be created if other materials, such as brick or concrete, had been used.

IM-2A: Types of Balance

- Use to show students that balance is the achievement of equilibrium in design.
- Have students sketch building designs that represent formal and informal types.

IM-2B Form Follows Function

- Use this drawing to show the different requirements of function and form in a variety of building types.

IM-2C Drafting Pencils

- This illustration shows the range of drafting leads available for board drafting. Ask students to identify the use of each.

EVALUATION

1. Assign the exercises at the end of the textbook chapter. When applicable, check students' answers against the answers given in the next section.

2. Test student mastery of factual material by using Chapter 2 Review found in the "Chapter Reviews" section of this guide.
3. In a check for competence, students should demonstrate understanding of:
   - the evolution of architectural styles and the people who developed modern styles.
   - the relationship of style to function and taste.
4. In a check for competence, students should be able to perform these skills:
   - recognize different architectural styles.
   - identify the ideas of influential architects.

ANSWERS TO END-OF-CHAPTER EXERCISES

Many end-of-chapter exercises ask students to perform an activity, such as draw a floor plan or photograph local structures. As a result, no "answer" is required and none is given here.

1. Answers will vary, but students should show awareness that form should be determined by use.
2. Answers will vary; students should show awareness of line, form, space, color, light and shadow, texture, and materials.
3. Answers will vary but should discuss variations in terms of the elements and principles of design.
4. Answers will vary.
5. Answers will vary; all six elements should be covered—line, form, space, color, light and shadow, texture, and materials.

Date	M	Tu	W	Th	F

Lesson Plan for Chapter 3 (Text pages 27–43)

Drafting Scales and Instruments

ESSENTIAL ELEMENTS

For a complete listing of the essential elements, see the "Program Development" section in this guide. Essential elements for the course: 3.4, 4.1, 4.2, 5.2, 6.9

CHAPTER 3 OUTLINE

Scales
- Architect's Scale
- Civil Engineer's Scale
- Metric Scales

Guides for Straight Lines
- T Square
- Parallel Slide
- Triangles
- Drafting Machine
- Protractor

Instruments for Curved Lines
- Compass
- Dividers
- Irregular Curve Instruments

Drafting and Lettering Tools
- Drafting Pencils
- Technical Pens

Papers and Drawing Surfaces

Correction Equipment

Timesaving Aids and Devices for Drafting
- Architectural Templates
- Overlays
- Tapes
- Stamps
- Underlays
- Squared Paper
- Burnishing Plates

OBJECTIVES

Following are the objectives that appear at the beginning of Chapter 3 in the textbook. Students are expected to learn to:

- measure and prepare drawings with different scales.
- draw with drafting instruments.
- select and use appropriate types of paper and other drafting supplies.
- use timesaving devices.

USING THE COLOR IMAGE MASTERS AND IMAGE MASTERS

The following color image masters (CIM), image masters (IM), and handout masters (HO) are appropriate for use with this chapter. Teaching suggestions are included.

IM-3A: Architect's Scales

- Point out that each inch is marked on the 1/4″ scale and each mark represents two inches on the 1/8″ scale because these are the usual limits of visual and instrumental precision.
- Ask students to determine what each division mark represents on the 1″, 1/2″, 3/4″, and 3/8″ scales.

HO-3A: Window Template

- Have students obtain a folder or notebook in which to keep the materials, such as this handout, that they will be given during the course.

- Point out to students that to use templates, the outline is followed inside the opening. On more complex templates, such as this one, different parts of the template are used for horizontal and vertical lines.
- Give students templates with which to experiment.

HO-3B: Making Geometric Calculations

- Architectural drawings involve a wide variety of geometric shapes. Stress to students that the ability to compute distances, areas, and volumes is important in estimating amounts of construction materials, labor, and costs. Give them these sheets to keep in their notebooks. They can refer to them when designing and working problems.

IM-3B Lettering

- Show this illustration to explain the basic rules for lettering.

EVALUATION

1. Assign the exercises at the end of the textbook chapter. When applicable, check students' answers against the answers given in the next section.
2. Test student mastery of factual material by using Chapter 3 Review found in the "Chapter Reviews" section of this guide.
3. In a check for competence, students should demonstrate understanding of:
   - how different types of scales work, including full size, fully divided, and open divided.
   - how to scale a selection to fit a drawing format.
   - the metric prefixes, base units, and scale selection.
   - how to read the vernier scale on a drafting machine.
   - differences in instrument quality and prices.
   - different types of templates and line tapes.
   - how using photographs in place of drawings reduces drawing time.
   - when the use of different types of timesaving devices is appropriate.
4. In a check for competence, students should be able to perform these skills:
   - measure scaled drawings using different architect's and civil engineer's scales.
   - measure full-size and scaled drawings using a metric scale.
   - draw using the T square, triangles, parallel slide, drafting machines, dividers, compass, beam compass, and templates.
   - use underlays and overlays.
   - draw with squared and isometric graph paper and on perspective grids.
   - construct a floor plan using tape.

ANSWERS TO END-OF-CHAPTER EXERCISES

Many end-of-chapter exercises ask students to perform an activity, such as draw a floor plan or photograph local structures. As a result, no "answer" is required and none is given here.

1. Lines should be drawn to the following lengths:
   a. 1 1/4″
   b. 1 7/8″
   c. 2 7/16″
   d. 2 13/16″
2. a. 5/8″
   b. 1 15/16″
   c. 1 7/32″
   d. 1 13/32″
3. No answer required.
4. Answers will vary.
5. 5′-6″ = 1676.4 mm
   6′-8″ = 2032 mm
   10′-4″ = 3149.6 mm
   11′-7″ = 3530.6 mm
   15′-3″ = 4648.2 mm
6. Answers will vary.
7. No answer required.
8. No answer required.
9. Answers will vary.

Date	M	Tu	W	Th	F

Lesson Plan for Chapter 4 (Text pages 44–58)

Architectural Drafting Conventions

ESSENTIAL ELEMENTS

For a complete listing of the essential elements, see the "Program Development" section in this guide. Essential elements for the course: 4.1, 5.3, 6.1, 6.2, 6.5, 6.9, 6.13

CHAPTER 4 OUTLINE

Architectural Drawings
- Types of Drawings
- Drawings and Documents
- Reading Architectural Drawings

Architectural Conventions
- Architectural Line Conventions
- Architectural Lettering

Architectural Drawing Techniques
- Using the Appropriate Pencil
- Rendering and Sketching Techniques

OBJECTIVES

Following are the objectives that appear at the beginning of Chapter 4 in the textbook. Students are expected to learn to:

- differentiate between the types and purposes of architectural drawings.
- produce the line conventions used on architectural drawings.
- develop good lettering techniques.
- sketch lines, patterns, and a floor plan.

USING THE COLOR IMAGE MASTERS AND IMAGE MASTERS

The following color image masters (CIM), image masters (IM), and handout masters (HO) are appropriate for use with this chapter. Teaching suggestions are included.

IM-4A: Drafting Materials

- Point out that drafting media are available in a wide variety of materials and sizes. Vellum facilitates printing, and its transparent surface aids tracing and alterations. Most residential drawings are done on size D sheets; E and F sizes are used for site drawings and large commercial and industrial drawings. Smaller sheets are used for preliminary detail drawings.

IM-4B: Line Thicknesses

- Use to illustrate the end result of different line thicknesses and their application.
- Have students try each of the leads mentioned in order to get a feel for them.

HO-4A: Lettering Styles

- Point out the example on the bottom of a lettering style that is difficult to read and should be avoided.
- Have students practice and develop basic skills in single-stroke Gothic lettering before allowing them to use more individual styles.

HO-4B: AIA Coding System

- Suggest that students keep this breakdown of AIA codes in their notebooks for easy reference.

IM-4C: Line Types

- Use this drawing to describe the application of basic line types.

HO-4C ASME Line Conventions

- This illustration summarizes the use of ASME line conventions.

EVALUATION

1. Assign the exercises at the end of the textbook chapter. When applicable, check students' answers against the answers given in the following section.
2. Test student mastery of factual material by using Chapter 4 Review found in the "Chapter Reviews" section of this guide.
3. In a check for competence, students should demonstrate understanding of:
   - different pencil lead grades.
   - architectural line conventions.
   - how to select lettering styles and the importance of clear lettering on a drawing.
4. In a check for competence, students should be able to perform these skills:
   - sketch all the different line weights.
   - create vertical, slant, or single-stroke Gothic lettering.
   - letter in an architectural style.

ANSWERS TO END-OF-CHAPTER EXERCISES

Many end-of-chapter exercises ask students to perform an activity, such as draw a floor plan or photograph local structures. As a result, no "answer" is required and none is given here.

1. Plans are top views; elevations show vertical views; sections show a slice through a structure; detail drawings give more precise information; renderings show how the finished product will look; and models are three-dimensional replicas.
2. General purpose drawings give approximate dimensions and are used for planning; working drawings contain all the information needed to construct the building.
3. Coding makes it easier to find and use a large number of drawings. Because all details could not be shown on one drawing, cross referencing guides the reader to other drawings having additional information.

4–11. No answers required.

Date	M	Tu	W	Th	F

Lesson Plan for Chapter 5 (Text pages 59–78)

Introduction to Computer-Aided Drafting and Design

ESSENTIAL ELEMENTS

For a complete listing of the essential elements, see the "Program Development" section in this guide. Essential elements for the course: 2.1, 3.2, 3.3, 3.4, 3.5, 4.1, 5.3, 6.5, 6.7, 6.13, 6.14

CHAPTER 5 OUTLINE

Advantages of CAD
CAD Limitations
Components of a CAD System
CAD Hardware
- The CPU
- The Monitor
- Memory and Storage
- Input Devices
- Output Devices
- Modems
- Multimedia Computers

CAD Software
Types of CAD Drawings
- Two-Dimensional Drawings
- Three-Dimensional Drawings

Using a CAD System
- Drawing Commands
- Editing Commands
- Utility Commands

Architectural Applications
- Steps in Creating a Floor Plan
- Steps in Creating an Elevation

OBJECTIVES

Following are the objectives that appear at the beginning of Chapter 5 in the textbook. Students are expected to learn:

- to use a computer to prepare architectural drawings.
- the different kinds of hardware and their functions.
- to evaluate CAD software programs.

USING THE COLOR IMAGE MASTERS AND IMAGE MASTERS

The following color image masters (CIM), image masters (IM), and handout masters (HO) are appropriate for use with this chapter. Teaching suggestions are included.

CIM-5A: Computer-Generated Pictorial

- This computer-generated pictorial was created by plotting and connecting points on contour lines, using X, Y, and Z coordinates on a perspective grid. Ask students to name the directions represented by X, Y, and Z.
- Surface treatments were added using library symbols. Ask students what other symbols were probably used.

EVALUATION

1. Assign the exercises at the end of the textbook chapter. When applicable, check students' answers against the answers given in the following section.
2. Test student mastery of factual material by using Chapter 5 Review found in the "Chapter Reviews" section of this guide.
3. In a check for competence, students should demonstrate understanding of:
   - how each component in a CAD system functions.
   - the advantages and disadvantages of using CAD.
   - the sequence for creating a basic architectural drawing using CAD.
   - how to make revisions on a CAD drawing.
4. In a check for competence, students should be able to perform these skills:
   - locate points along X and Y axes using grid paper.
   - select the appropriate CAD tool for different drafting tasks.
   - locate and position symbols from CAD libraries.
   - activate CAD commands.
   - create floor plans, details, and 3-D drawings using CAD.

ANSWERS TO END-OF-CHAPTER EXERCISES

Many end-of-chapter exercises ask students to perform an activity, such as draw a floor plan or photograph local structures. As a result, no "answer" is required and none is given here.

1. Monitor, CPU, microprocessor, graphic accelerator, CD-ROM drive, hard disk drive, floppy disk drives, tape drive, keyboard, digitizer, puck, mouse.
2. CAD makes revisions faster and cleaner, automates repetitious tasks, increases accuracy and consistency. Because drawings are easier to do, more are drawn, which means fewer construction mistakes.
3. People create drawings, not computers; machines can't think or make decisions or eliminate errors.
4. No answer required.
5. No answer required.
6. No answer required.
7. Draw outer walls; draw inner walls; insert windows; insert doors; insert kitchen and bath fixtures; add notes and labels; dimension the drawing.
8. No answer required.

Date	M	Tu	W	Th	F

Lesson Plan for Chapter 6 (Text pages 79–94)

Environmental Design Factors

ESSENTIAL ELEMENTS

For a complete listing of the essential elements, see the "Program Development" section in this guide. Essential elements for the course: 2.2, 2.3, 2.4, 3.3, 3.5, 4.4, 5.3, 6.7, 6.10

CHAPTER 6 OUTLINE

Orientation
- Energy Orientation and Sources
- Solar Orientation
- Land and a Structure
- Wind Control

Ergonomic Planning
- Human Dimensions
- Safety Factors

Ecology
- Land Pollution
- Air Pollution
- Water Pollution
- Visual Pollution
- Noise Pollution
- Electronic Hazards
- Preventive Measures

OBJECTIVES

Following are the objectives that appear at the beginning of Chapter 6 in the textbook. Students are expected to learn:

- to orient a house on a lot to take best advantage of solar energy and features of the lot.
- to design structures ergonomically.
- ways to prevent pollution (ecology).

USING THE COLOR IMAGE MASTERS AND IMAGE MASTERS

The following color image masters (CIM), image masters (IM), and handout masters (HO) are appropriate for use with this chapter. Teaching suggestions are included.

CIM-6A: Stone Residence Site

- Show this illustration as an example of a residence designed to integrate with nature without destroying natural contours and vegetation.
- Ask students to tell in what ways the natural vegetation will influence sun and wind exposure.

IM-6A: Seasonal Shadows

- Point out that shadow is an important factor in planning building orientation. Sun angles during different seasons can help determine what areas will be shielded or in full sun during different parts of the day. The computer-generated shadow patterns in the illustration were done to show what buildings would be sun blocked by high-rises for a New York City project.
- Ask students to select a tree or tall structure visible from the school and note the positions of its shadow at 9 AM, 2 PM, and 3 PM.

HO-6A: Environmental Design

- This handout is a six page summary of the principles of designing an environmentally sound house.

HO-6B: Toxic Enigma

- This handout consists of an eleven page introduction to toxins and their control in the architectural design process.

HO-6C: ADA

- This handout describes the portions of the Americans with Disabilities Act which apply to the design of structures to be used by persons with disabilities.

EVALUATION

1. Assign the exercises at the end of the textbook chapter. When applicable, check students' answers against the answers given in the following section.
2. Test student mastery of factual material by using Chapter 6 Review found in the "Chapter Reviews" section of this guide.
3. In a check for competence, students should demonstrate understanding of:
   - problems resulting from pollution.
   - effects of air, water, land, sound, and visual pollution on their communities.
4. In a check for competence, students should be able to perform these skills:
   - list environmental factors involved in protecting a building site.
   - list possible ways to stop air, water, land, sound, and visual pollution.

ANSWERS TO END-OF-CHAPTER EXERCISES

Many end-of-chapter exercises ask students to perform an activity, such as draw a floor plan or photograph local structures. As a result, no "answer" is required and none is given here.

1. The south and west sides of a structure are the warmest and brightest because of almost constant exposure to the sun. The east and north sides are the coolest and darkest because they do not receive as much sunlight. In the winter, the angle of the sun is low; in the summer the sun is higher in the sky.
2. Drawings will vary. To take advantage of solar orientation, the kitchen and dining areas should face the south or east; the living room should face south or west; the bedrooms should face north.
3. No answer required.
4. No answer required.
5. Answers will vary.
6. No answer required.
7. Answers will vary, depending upon the buildings identified.
8. Answers will vary, depending upon the type of home desired.
9. No answer required (see Fig. 6-28).

Date	M	Tu	W	Th	F

Lesson Plan for Chapter 7 (Text pages 95–118)

Indoor Living Areas

ESSENTIAL ELEMENTS

For a complete listing of the essential elements, see the "Program Development" section in this guide. Essential elements for the course: 2.2, 3.1, 3.5, 5.3, 6.6, 6.7, 6.12

CHAPTER 7 OUTLINE

Living Area Plans
- Open Plan
- Closed Plan
- Combined Plans

Living Rooms
- Function
- Location
- Orientation
- Decor
- Size and Shape

Dining Rooms
- Function and Location
- Decor
- Size and Shape

Family Rooms
- Function
- Location
- Decor
- Size and Shape

Recreation Rooms
- Function
- Location
- Decor
- Size and Shape

Special-Purpose Rooms
- Function
- Location
- Decor and Lighting
- Size and Shape

OBJECTIVES

Following are the objectives that appear at the beginning of Chapter 7 in the textbook. Students are expected to learn:

- to identify the functions of indoor living area rooms.
- to design the location, decor, size, and shape of indoor living areas.
- how a room's orientation, walls, floors, windows, ceilings, lighting, and furniture can contribute to room function and appearance.
- to design indoor living areas and work them into a convenient floor plan.

USING THE COLOR IMAGE MASTERS AND IMAGE MASTERS

The following color image masters (CIM), image masters (IM), and handout masters (HO) are appropriate for use with this chapter. Teaching suggestions are included.

CIM-7A: Living Room

- This is the southwest living room wall of the plan shown in Fig. 14-20 and in the set of plans accompanying this instructor's guide. Using the plans, have students identify the facing walls and indicate which rooms are behind each one.
- Ask students how structural features have been used as design elements in this room.

CIM-7B: Fireplace

- Because gas fireplaces do not require large flues directly above the firebox, there can be an open space between the fireplace and the ceiling. Ask students what effect is created by the half-wall fireplace in this picture.

- Ask what elements and principles of design have been used in this fireplace.

CIM-7C: Living Area

- Ask students if this shows an open or closed plan. What criteria did they use to decide?
- Ask what function the fireplace has besides being a heat source.

HO-7A: Checklist for Prospective Homeowners

- Give to students to keep in their notebooks as a handy guide to planning or purchasing a home.

EVALUATION

1. Assign the exercises at the end of the textbook chapter. When applicable, check students' answers against the answers given in the following section.
2. Test student mastery of factual material by using Chapter 7 Review found in the "Chapter Reviews" section of this guide.
3. In a check for competence, students should demonstrate understanding of:
   - the functions and basic decor of living areas.
   - how to locate living areas.
   - the functions and locations of dining areas.
   - types of dining room furniture and their space requirements.
   - the functions and locations of a family room.
   - types of storage in a family room.
   - the functions and locations of a den.
   - types of games that take place in a recreation room.
   - the locations for a recreation room and types of storage.
4. In a check for competence, students should be able to perform these skills:
   - sketch open- and closed-plan living areas and make a list of furnishings for them.
   - design a living area around furniture templates.
   - sketch both large and small dining rooms having table and chairs.
   - design formal, informal, and outdoor dining rooms.
   - sketch both a large and a small family room and add furnishings.
   - sketch both a large and a small den and add furnishings.
   - sketch both a large and a small recreation room and add furniture and game equipment.

ANSWERS TO END-OF-CHAPTER EXERCISES

Many end-of-chapter exercises ask students to perform an activity, such as draw a floor plan or photograph local structures. As a result, no "answer" is required and none is given here.

1–9. No answer required.

10. 10′ × 12′

11. 12′ × 17′-6″; answers will vary

12–14. Answers will vary.

15–18. No answer required.

Date	M	Tu	W	Th	F

Lesson Plan for Chapter 8 (Text pages 119–132)

Outdoor Living Areas

ESSENTIAL ELEMENTS

For a complete listing of the essential elements, see the "Program Development" section in this guide. Essential elements for the course: 1.1, 2.2, 3.1, 3.5, 5.2, 5.3, 6.6, 6.7, 6.12

CHAPTER 8 OUTLINE

Porches
- Function and Types
- Location
- Decor
- Size and Shape

Patios
- Function and Types
- Location
- Decor
- Size and Shape

Lanais
- Function
- Location
- Decor
- Size and Shape

Swimming Pools
- Function
- Location and Orientation
- Pool Construction
- Safety Devices
- Pool Equipment

OBJECTIVES

Following are the objectives that appear at the beginning of Chapter 8 in the textbook. Students are expected to learn to:

- design and sketch a porch, patio, and lanai.
- design and sketch a swimming pool.
- calculate the area and volume of swimming pools.

USING THE COLOR IMAGE MASTERS AND IMAGE MASTERS

The following color image masters (CIM), image masters (IM), and handout masters (HO) are appropriate for use with this chapter. Teaching suggestions are included.

CIM-8A: Pool

- Have students turn to Fig. 18-28 in their texts to see the plan for this pool. It is also shown in the set of drawings that accompany this teacher's resource binder.
- Ask students to calculate the pool's volume based on dimensions given in Fig. 18-28 in the text.

IM-8A: Pool Shapes

- Point out that pool shapes were once limited to very simple rectangles unless expensive construction methods were used. Now almost any shape and depth can be made using concrete blown into molds and reinforced with steel.
- Ask students to create three freeform pool designs—one for a family with children, one for a retired couple, and one for an apartment catering to singles. Ask them to give overall dimensions and state their reasons for the different designs.

EVALUATION

1. Assign the exercises at the end of the textbook chapter. When applicable, check students' answers against the answers given in the following section.

2. Test student mastery of factual material by using Chapter 8 Review found in the "Chapter Reviews" section of this guide.
3. In a check for competence, students should demonstrate understanding of:
   - the different styles, functions, and locations of a patio.
   - the types of materials used to build patios.
   - methods used to protect patios from the weather.
4. In a check for competence, students should be able to perform these skills:
   - sketch both a large and small patio.
   - sketch both a secluded and a dining patio.

ANSWERS TO END-OF-CHAPTER EXERCISES

Many end-of-chapter exercises ask students to perform an activity, such as draw a floor plan or photograph local structures. As a result, no "answer" is required and none is given here.

1–6. No answers required.

7. Serves as an exterior hallway; answers will vary.

8. Water pump, filters, purifiers, skimmer, heater, pipes, outlets. Optional: diving board, spa, screened enclosure, fountain.

9. Building codes, relationship with the house, sun exposure, privacy, controlled access, materials used in construction, shape and size, safety devices, maintenance equipment.

10–11. No answers required.

Date	M	Tu	W	Th	F

Lesson Plan for Chapter 9 (Text pages 133–145)

Traffic Areas and Patterns

ESSENTIAL ELEMENTS

For a complete listing of the essential elements, see the "Program Development" section in this guide. Essential elements for the course: 2.2, 3.1, 3.5, 5.3, 6.6, 6.7, 6.12

CHAPTER 9 OUTLINE

Traffic Patterns
Halls
Stairs
    Materials and Lighting
    Size and Shape
Entrances
    Function and Types
    Location
    Decor
    Size and Shape

OBJECTIVES

Following are the objectives that appear at the beginning of Chapter 9 in the textbook. Students are expected to learn:

- to determine the effectiveness of a traffic pattern in a house.
- to plan hallways that function efficiently.
- guidelines for designing stairs.
- to calculate the correct space needed for stairways and stairwells.
- the kinds and functions of entrances.
- guidelines for entrance design.
- to design a foyer and entry.

USING THE COLOR IMAGE MASTERS AND IMAGE MASTERS

The following color image masters (CIM), image masters (IM), and handout masters (HO) are appropriate for use with this chapter. Teaching suggestions are included.

CIM-9A: Driveway

- Ask students how this driveway provides for parking and the continuous flow of traffic.
- Ask in what ways the design of this entrance is consistent with the overall design of the structure.

IM-9A: Foyer Layout

- This drawing shows a typical foyer plan as a reference in floor planning.

EVALUATION

1. Assign the exercises at the end of the textbook chapter. When applicable, check students' answers against the answers given in the following section.
2. Test student mastery of factual material by using Chapter 9 Review found in the "Chapter Reviews" section of this binder.
3. In a check for competence, students should demonstrate understanding of:
   - principles used to create efficient halls and stairs.
   - the relationship of an entry to the traffic pattern in a structure.
4. In a check for competence, students should be able to perform these skills:
   - sketch large, small, and two-story floor plans having efficient traffic patterns.

ANSWERS TO END-OF-CHAPTER EXERCISES

Many end-of-chapter exercises ask students to perform an activity, such as draw a floor plan or photograph local structures. As a result, no "answer" is required and none is given here.

1–3. No answers required.

4. L shape, double L (U) shape, spiral, straight run.
5. Site entrance gives entry from road and a place to park; main house entrance gives access to house and a waiting area; service entrance is used when main entrance would be inappropriate, such as for deliveries; special-purpose entrance gives access to outside living areas.

6–9. No answers required.

Date	M	Tu	W	Th	F

Lesson Plan for Chapter 10 (Text pages 146–159)
Kitchens

ESSENTIAL ELEMENTS

For a complete listing of the essential elements, see the "Program Development" section in this guide. Essential elements for the course: 2.2, 3.1, 3.5, 5.3, 6.6, 6.7, 6.12

CHAPTER 10 OUTLINE

Kitchen Design Considerations
- Functions
- Types of Kitchens
- Decor
- Size and Shape

Kitchen Planning Guidelines

OBJECTIVES

Following are the objectives that appear at the beginning of Chapter 10 in the textbook. Students are expected to learn to:

- apply guidelines to efficient kitchen design.
- determine the best shape, size, and location for the kitchen.
- plan and draw a work triangle for a kitchen.
- design an aesthetically consistent decor for a kitchen.
- sketch small and large kitchens of the basic kitchen shapes.

USING THE COLOR IMAGE MASTERS AND IMAGE MASTERS

The following color image masters (CIM), image masters (IM), and handout masters (HO) are appropriate for use with this chapter. Teaching suggestions are included.

CIM-10A: **Island Kitchen**

- Ask students what purposes the island serves in this kitchen.
- Point out the large window behind the sink area and the use of molding on the cabinets and appliances.
- Ask students to identify the major activities areas.

CIM-10B: **Peninsula Kitchen**

- This kitchen has two peninsulas, one at a higher level and one at a lower level. Ask students to identify the tasks done at each.
- There is also a separate full-size sink on the left wall and a range on the rear wall. Ask students to draw a work triangle for this kitchen.

IM-10A: **Kitchen Layout**

- Use this drawing to show a kitchen plan with a counter designed for eating and food preparation.

EVALUATION

1. Assign the exercises at the end of the textbook chapter. When applicable, check students' answers against the answers given in the following section.
2. Test student mastery of factual material by using Chapter 10 Review found in the "Chapter Reviews" section of this binder.
3. In a check for competence, students should demonstrate understanding of:
   - the functions and locations of a kitchen.
   - the working areas of a kitchen and the principle of the work triangle.
   - wall and base cabinet dimensions and heights of working surfaces.

4. In a check for competence, students should be able to perform these skills:
   - sketch both small and large kitchens using the six basic kitchen shapes.
   - sketch a large family kitchen and its furnishings.

ANSWERS TO END-OF-CHAPTER EXERCISES

Many end-of-chapter exercises ask students to perform an activity, such as draw a floor plan or photograph local structures. As a result, no "answer" is required and none is given here.

1. *U-shaped*—advantages: efficient, traffic outside work triangle; disadvantage: doors and drawers must be designed to open without interfering with one another.

   Peninsula—advantage: large work areas; disadvantage: cabinet space may be limited.

   L-shaped—advantages: traffic outside work triangle, requires less space than U-shaped; disadvantage: if the walls are too long, compactness is destroyed.

   Corridor—advantages: efficient for narrow rooms and limited space, efficient work triangle; disadvantage: traffic passing through work triangle destroys efficiency.

   One-wall—advantage: efficient for limited spaces; disadvantages: walls that are too long can destroy efficiency, storage often limited.

   Island—advantages: convenient for multiple workers, island accessible from all sides; disadvantage: requires a large space.

   Family—advantages: provides meeting place for family; disadvantage: requires a large space.

2–6. No answers required.

7. Answers will vary.

8. No answer required.

Date	M	Tu	W	Th	F

Lesson Plan for Chapter 11 (Text pages 160–174)

General Service Areas

ESSENTIAL ELEMENTS

For a complete listing of the essential elements, see the "Program Development" section in this guide. Essential elements for the course: 2.2, 3.1, 3.5, 5.3, 6.6, 6.7, 6.12

CHAPTER 11 OUTLINE

Utility Rooms
- Function
- Location
- Style and Decor
- Size and Shape

Garages and Carports
- Function and Location
- Decor
- Size

Driveways

Workshops
- Function and Location
- Decor
- Size

Storage Areas
- Function and Types
- Location

OBJECTIVES

Following are the objectives that appear at the beginning of Chapter 11 in the textbook. Students are expected to learn to:

- determine what kinds of equipment are included in a utility room.
- evaluate the best location for a utility room.
- sketch a garage and a carport.
- design storage facilities for a garage.
- calculate the area needed for garages and driveways.
- design and sketch an efficient and safe workshop area.
- design and sketch service area storage facilities.

USING THE COLOR IMAGE MASTERS AND IMAGE MASTERS

The following color image masters (CIM), image masters (IM), and handout masters (HO) are appropriate for use with this chapter. Teaching suggestions are included.

IM-11A: Garage Roofs

- Use this to illustrate how the appearance of a garage can be altered by means of roof and door changes. (These renderings were computer generated using ACCURENDER.)
- Ask students how the elements and principles of design play a part in each garage treatment.

IM-11B: Garage Storage

- Use to show how otherwise wasted space over a car hood can be utilized. Other over-the-hood space savers include workbenches and storage cabinets.
- Ask students to suggest other over-the-hood ideas.

HO-11A: Auto Dimensions

- Suggest that students refer to this handout when planning garage space. First they should use a car template and then add the necessary space all around.

IM-11C Garage Living Combo

- Point out how living space can be incorporated into a garage design using this drawing.

EVALUATION

1. Assign the exercises at the end of the textbook chapter. When applicable, check students' answers against the answers given in the following section.
2. Test student mastery of factual material by using Chapter 11 Review found in the "Chapter Reviews" section of this binder.
3. In a check for competence, students should demonstrate understanding of:
   - the function, locations, and working order of a utility room.
   - the functions and locations of home work areas, the different types of workbenches and storage, and the types of hand and power tools required.
   - the functions of a garage and carport, types of garage doors and driveways, and the locations for parking.
4. In a check for competence, students should be able to perform these skills:
   - sketch both small and large utility rooms, including all fixtures and appliances.
   - sketch both large and small work areas, in both home and garage, which include storage and power tools.
   - sketch single and double garages, including autos, storage, and utility areas.

ANSWERS TO END-OF-CHAPTER EXERCISES

Many end-of-chapter exercises ask students to perform an activity, such as draw a floor plan or photograph local structures. As a result, no "answer" is required and none is given here.

1–7. No answers required.

Date	M	Tu	W	Th	F

Lesson Plan for Chapter 12 (Text pages 175–192)

Sleeping Areas

ESSENTIAL ELEMENTS

For a complete listing of the essential elements, see the "Program Development" section in this guide. Essential elements for the course: 2.2, 3.1, 3.5, 5.3, 6.6, 6.7, 6.12

CHAPTER 12 OUTLINE

Bedrooms
- Function
- Location
- Decor
- Size and Shape

Baths
- Function
- Location
- Decor
- Size and Shape

OBJECTIVES

Following are the objectives that appear at the beginning of Chapter 12 in the textbook. Students are expected to learn to:

- plan and draw bedrooms for a sleeping area.
- plan and draw baths appropriate to the size and the arrangement of the floor plan.
- design an efficient bath.

USING THE COLOR IMAGE MASTERS AND IMAGE MASTERS

The following color image masters (CIM), image masters (IM), and handout masters (HO) are appropriate for use with this chapter. Teaching suggestions are included.

CIM-12A: Bathroom

- This bath has an integrated shower and whirlpool tub with built-in ledges for entry and drying off. Ask students how privacy is maintained. If neighbors are too near, what else could enhance privacy besides the use of blinds?
- What other unique features are present in this design?

IM-12A: Master Bedroom and Study

- Ask students how the make-up area could be enlarged and separated from the bedroom in this compartment master bedroom suite. How could more storage space be added to the closet?
- Ask students to redesign this suite using the existing space.

HO-12A: Closet and Bathroom Spacing

- Have students use this handout when designing closet and bathroom space.

HO-12B: Master Bedroom Suite

- Have students add details to closets and eliminate furniture offsets by designing built-in facilities.

EVALUATION

1. Assign the exercises at the end of the textbook chapter. When applicable, check students' answers against the answers given in the following section.
2. Test student mastery of factual material by using Chapter 12 Review found in the "Chapter Reviews" section of this binder.
3. In a check for competence, students should demonstrate understanding of:
   - the functions, locations, ventilation, and space requirements of bedrooms.

- how to reduce noise in the bedroom.
- the functions and locations of bathrooms, including space requirements for fixtures.

4. In a check for competence, students should be able to perform these skills:
   - sketch small and medium-sized bedrooms with furnishings.
   - sketch a master bedroom with furnishings that includes a private bath and dressing room.
   - sketch a full bathroom, half bath, three-quarter bath (with shower) and partitioned bath.

ANSWERS TO END-OF-CHAPTER EXERCISES

Many end-of-chapter exercises ask students to perform an activity, such as draw a floor plan or photograph local structures. As a result, no "answer" is required and none is given here.

1–5. No answers required.

6. Proportional dimensions may vary depending on door and window placement and furniture orientation (10′ × 12′ min.).

7–11. No answers required.

12. Answers will vary.

Date	M	Tu	W	Th	F

Lesson Plan for Chapter 13 (Text pages 193–218)
Designing Floor Plans

ESSENTIAL ELEMENTS

For a complete listing of the essential elements, see the "Program Development" section in this guide. Essential elements for the course: 3.1, 3.3, 3.5, 4.3, 5.3, 6.7, 6.10, 6.15

CHAPTER 13 OUTLINE

Floor Plan Development
The Design Process
- Defining the Project
- Analyzing the Project
- Developing a Conceptual Design
- Evaluating the Design
- Design Development

Functional Space Planning
- Planning Space for Rooms and Areas
- Floor Plan Sketches

Developing Plans to Accommodate Special Needs
- Public Buildings
- Residences

OBJECTIVES

Following are the objectives that appear at the beginning of Chapter 13 in the textbook. Students are expected to learn to:

- gather information from a client that is needed to design an architectural project.
- analyze a building site.
- use the design process to prepare for drawing accurate and functional floor plans.
- create floor plan sketches.
- design floor plans to accommodate the needs of persons with physical impairments.

USING THE COLOR IMAGE MASTERS AND IMAGE MASTERS

The following color image masters (CIM), image masters (IM), and handout masters (HO) are appropriate for use with this chapter. Teaching suggestions are included.

CIM-13A: Floor Plan

- This computer-generated floor plan has layers of different colors to identify different functions and components. Have students list the colors used and what they represent.
- Ask students to list the advantages of using a computer to draw this plan.

IM-13A: Cabin Floor Plan

- Ask students to redesign and sketch the allocation of living, kitchen, and bath space to meet their own personal needs.

IM-13B: Single-Line Floor Plan

- Ask students to redesign and sketch this layout to create a foyer and eliminate offset areas, such as the utility room.

HO-13A: Arranging Geometric Space in Floor Plans

- Discuss with students the different ways in which geometric space can be arranged to accommodate different house plans.

HO-13B: Checklists for Architectural Working Drawings

- Give to students to keep in their notebooks and use to determine if any appropriate item has been omitted from a particular drawing. Not all drawings must include each item on the list. However, it is good practice for students to think about whether a symbol, entry, dimension, or note is needed.

- The instructor may use the list in checking and grading drawings.

HO-13C: Barrier-Free Residential Design

- This handout expands on information in the textbook. Have students keep it in their notebooks for reference when designing for people with physical impairments.

IM-13C Floor Plan Patterns

- This illustration shows the ten most used plan configurations of areas used in residential design. Have students list the advantages and limitations of each.

HO-13D Residential Design

- For students interested in a more detailed description of the design process use this six page summary of residential design to reinforce and amplify the text coverage.

EVALUATION

1. Assign the exercises at the end of the textbook chapter. When applicable, check students' answers against the answers given in the following section.
2. Test student mastery of factual material by using Chapter 13 Review found in the "Chapter Reviews" section of this binder.
3. In a check for competence, students should demonstrate understanding of:
   - the elements and principles of design and the theory of color.
   - the architectural design process.
   - the need for communication between client and designer.
   - room size and the relationship of individual rooms to overall design.
   - the relationships among living, sleeping, and service areas.
   - open and closed plans and the expandable plan.
   - traffic patterns and how to check for them on a floor plan.
   - how to check dimensions on a floor plan.
4. In a check for competence, students should be able to perform these skills:
   - redesign and sketch a small floor plan and make revisions.
   - list the steps in the design sequence.
   - identify wants and needs in terms of design.
   - select colors appropriate for a desired atmosphere.
   - design a floor plan and make furniture templates for it.
   - select the right furnishings of the correct size and design rooms to fit furnishing requirements.
   - sketch, refine, and finalize a complete floor plan.
   - check a floor plan with furniture templates and make new templates for specific sizes of furniture.

ANSWERS TO END-OF-CHAPTER EXERCISES

Many end-of-chapter exercises ask students to perform an activity, such as draw a floor plan or photograph local structures. As a result, no "answer" is required and none is given here.

1. Define the project, analyze the project; develop a conceptual design; evaluate the design; develop the final design.
2. Answers will vary.
3. The soil analysis drawing is placed over the slope analysis. Property lines are aligned with the base map. Distinct areas are traced on an overlay. The overlay is then placed over the visual analysis drawing and areas are again outlined. This helps determine the best locations for structures on the site.

4–6. No answers required.

7. Answers will vary.

8–11. No answers required.

Date	M	Tu	W	Th	F

Lesson Plan for Chapter 14 (Text pages 219–242)

Drawing Floor Plans

ESSENTIAL ELEMENTS

For a complete listing of the essential elements, see the "Program Development" section in this guide. Essential elements for the course: 3.5, 5.2, 5.3, 6.2, 6.5, 6.6, 6.7, 6.9, 6.13, 6.15

CHAPTER 14 OUTLINE

Types of Floor Plans
Floor-Plan Symbols
    Wall Symbols
    Door Symbols
    Window Symbols
    Appliance and Fixture Symbols
    Bath Symbols
    Furniture Symbols
Steps in Drawing Floor Plans
    Size and Scale in Floor Plans
Multiple-Level Floor Plans
    Drawing Separate Plans
    Calculating Dimensions for Stair Systems
Reversed Plans
Reflected Ceiling Plans
Floor-Plan Dimensioning
    Drawing Dimensions
    Providing Dimensions
Guidelines for Dimensioning
    List of Guidelines
CAD Dimensioning
Metric Dimensioning
Modular Dimensions

OBJECTIVES

Following are the objectives that appear at the beginning of Chapter 14 in the textbook. Students are expected to learn to:

- use information on a scaled floor plan to draw a complete floor plan.
- name and explain the types of floor plans.
- use graphic symbols to communicate information on a floor plan.
- draw a floor plan according to a sequence of steps.
- draw dimensions that convey precise, accurate information for builders.

USING THE COLOR IMAGE MASTERS AND IMAGE MASTERS

The following color image masters (CIM), image masters (IM), and handout masters (HO) are appropriate for use with this chapter. Teaching suggestions are included.

IM-14A: Sequential Development of Floor Plan

- Review the steps in making a floor plan from layout to outlining to adding symbols and dimensions.
- Ask students to identify the lines used in step 4.

IM-14B: Living Room Views

- Use to show how different views from the living room were considered for the house shown in Fig. 14-20 in the text and in the set of plans that accompany this instructor's guide.
- Have students draw view lines on a floor plan overlay to verify or alter the design.

HO-14A: Master Bath

- This alternate master bath was drawn for the home shown in the set of drawings that accompanies this instructor's guide. Have students check the design that was actually used and compare the two plans.

- Have students sketch a third alternative.

HO-14B: Reflected Ceiling Plan

- Point out to students that a reflected ceiling plan is usually prepared for buildings with ceiling designs involving multiple lighting fixtures or levels. The ceiling plan has the same shape and dimensions as the floor plan, as if the floor were a mirror.

HO-14C: Floor Plan Symbols for Common Doors

- Have students keep this handout in their notebooks for use when creating floor plans.

HO-14D: Floor Plan Symbols for Common Windows

- Have students keep this handout in their notebooks for use when creating floor plans.

HO-14E: Floor Plan Symbols for Appliances and Fixtures

- Have students keep this handout in their notebooks for use when creating floor plans.

HO-14F: Floor Plan Symbols for Plumbing Facilities

- Have students keep this handout in their notebooks for use when creating floor plans.

IM-14C: Floor Plan and Elevations

- Use this illustration to show the relationship between floor plans and both interior and exterior elevations.

IM-14D: Floor Plan Dimensions

- This plan is an example of standard floor plan dimensioning methods.

IM-14E: Floor Plan Symbols

- This drawing shows the appropriate scale to be used for floor plan symbols.

IM-14F: Metric Dimensions

- This drawing illustrates accepted methods of metric dimensioning.

EVALUATION

1. Assign the exercises at the end of the textbook chapter. When applicable, check students' answers against the answers given in the following section.
2. Test student mastery of factual material by using Chapter 14 Review found in the "Chapter Reviews" section of this guide.
3. In a check for competence, students should demonstrate understanding of:
   - the different floor plan symbols.
   - the correct sequence in which to draw floor plans.
   - the rules for floor plan dimensioning, the different methods used, and how to check dimensions.
   - modular dimensions.
4. In a check for competence, students should be able to perform these skills:
   - draw single and multilevel floor plans having all required symbols and dimensions.
   - draw a reversed floor plan.
   - draw different types of arrowheads.

ANSWERS TO END-OF-CHAPTER EXERCISES

Many end-of-chapter exercises ask students to perform an activity, such as draw a floor plan or photograph local structures. As a result, no "answer" is required and none is given here.

1–5. No answers required.

Date	M	Tu	W	Th	F

Lesson Plan for Chapter 15 (Text pages 243–257)

Designing Elevations

ESSENTIAL ELEMENTS

For a complete listing of the essential elements, see the "Program Development" section in this guide. Essential elements for the course: 2.4, 3.1, 3.3, 3.5, 5.3, 4.3, 6.4, 6.7, 6.9, 6.10

CHAPTER 15 OUTLINE

Relationship with the Floor Plan
Elements of Design and Elevations
  Lines
  Form and Space
  Light and Color
  Materials and Texture
Design Sequence for Creating Elevations
  Roof Style and Type
  Window Styles
  Door Styles

OBJECTIVES

Following are the objectives that appear at the beginning of Chapter 15 in the textbook. Students are expected to learn to:

- apply the principles and elements of design to creating elevation drawings.
- recognize different roof styles as options for roof design.
- select and design window styles in relation to elements of design and window functions.
- locate doors on an elevation design, considering style, size, and types of doors.

USING THE COLOR IMAGE MASTERS AND IMAGE MASTERS

The following color image masters (CIM), image masters (IM), and handout masters (HO) are appropriate for use with this chapter. Teaching suggestions are included.

CIM-15A: Florida Residence–Southwest Elevation

- This photo is of the house shown in the set of plans accompanying this instructor's guide. Have students compare the drawing with the photo of the finished house. Ask students to compare key positions in the picture with floor and site plan details. Are they identical?

CIM-15B: Florida Residence–Southeast Elevation

- This photo is of the house shown in the set of plans accompanying this teacher's resource binder. Have students compare the drawing with the photo of the finished house. Ask students to compare key positions in the picture with floor and site plan details. Are they identical?

IM-15A: True Lengths in Orthographic Drawings

- Explain that in orthographic drawings, true length is used to represent actual dimensions on the horizontal (x) and vertical (y) axes. However, when two points are at different distances from the plane of projection, the resulting lines are not true length when viewed from an orthographic plane. Instead, they are fore-shortened, like the roof lines in this illustration.

HO-15A: Roof Outlines

- Point out that, if a period style is to be authentic, the roof outline must be correct. This handout shows roof lines associated with the six Early American styles.

IM-15B Plan and Exterior Elevations

- This drawing shows the correct orientation of elevations with a floor plan.

IM-15C Roof Pitch

- Use this illustration to describe the terminology used to define roof pitches.

IM-15D Roof Pitch Calculation

- Show this illustration to students to illustrate the formula used to determine roof pitch.

IM-15E Roof Projected Dimensions

- The concept of foreshortening is difficult for some students. This drawing simplifies and shows the comparison of true and foreshortened dimensions of a roof.

IM-15F Pictorial and Elevations

- Proceeding from the familiar to the unfamiliar this illustration shows a comparison of a pictorial view with four elevation views.

EVALUATION

1. Assign the exercises at the end of the textbook chapter. When applicable, check students' answers against the answers given in the following section.
2. Test student mastery of factual material by using Chapter 15 Review found in the "Chapter Reviews" section of this binder.
3. In a check for competence, students should demonstrate understanding of:
   - the relationship of the elevation to the floor plan.
   - the relationship of the elevation to the roof plan.
   - basic elevation types.
4. In a check for competence, students should be able to perform these skills:
   - sketch elevations that have a pleasing appearance.
   - sketch elevations in which hip, gable, shed, mansard, gambrel, and flat roof designs are shown.

ANSWERS TO END-OF-CHAPTER EXERCISES

Many end-of-chapter exercises ask students to perform an activity, such as draw a floor plan or photograph local structures. As a result, no "answer" is required and none is given here.

1–5. No answers required.

Date	M	Tu	W	Th	F

Lesson Plan for Chapter 16 (Text pages 258–280)

Drawing Elevations

ESSENTIAL ELEMENTS

For a complete listing of the essential elements, see the "Program Development" section in this guide. Essential elements for the course: 3.5, 5.3, 6.2, 6.3, 6.5, 6.6, 6.7, 6.9, 6.15

CHAPTER 16 OUTLINE

Elevation Projection
- Purposes of Projected Elevation Drawings

Drawing Elevations from a Floor Plan
- Orientation
- Auxiliary Elevations
- Elevations and Construction
- Steps in Projecting Elevations

Elevation Symbols
- Material Symbols
- Window Symbols
- Door Symbols

Interior Elevations
- Steps in Drawing an Interior Elevation

Elevation Dimensioning
- Standards and Guidelines for Dimensioning
- Types of Dimensions
- Interior vs. Exterior Dimensioning

Landscape on Elevation Drawings
- Creating a Realistic Drawing
- Steps for Making Landscaped Elevations

Drawing and Rendering Techniques

OBJECTIVES

Following are the objectives that appear at the beginning of Chapter 16 in the textbook. Students are expected to learn:

- to follow a series of steps to project elevations from a floor plan and complete an elevation drawing.
- to draw accurately scaled and dimensioned elevations.
- to mathematically establish the pitch of a roof.
- symbols used on elevations.
- pictorial drawing and rendering techniques to use on elevations.

USING THE COLOR IMAGE MASTERS AND IMAGE MASTERS

The following color image masters (CIM), image masters (IM), and handout masters (HO) are appropriate for use with this chapter. Teaching suggestions are included.

IM-16A: Kitchen Elevation

- This is the outside kitchen wall shown in Fig. 14-20 and in the set of drawings that accompany this teacher's resource binder. Have students identify the position of windows on the floor plan and on the exterior elevation in the set of drawings.

IM-16B Bedroom Elevations

- This drawing shows the relationship between a floor plan and an interior elevation of a bedroom wall.

IM-16C Kitchen Elevations

- This drawing shows the relationship between a floor plan and interior elevations of a kitchen wall.

IM-16D Bath Elevations

- This drawing shows the relationship between a floor plan and interior elevations of a bath wall.

IM-16E Office Elevations

- This drawing shows the relationship between a floor plan and interior elevations of an office wall.

EVALUATION

1. Assign the exercises at the end of the textbook chapter. When applicable, check students' answers against the answers given in the following section.
2. Test student mastery of factual material by using Chapter 16 Review found in the "Chapter Reviews" section of this guide.
3. In a check for competence, students should demonstrate understanding of:
   - projection of the floor plan to the elevation drawing and the relationship between the symbols on each.
   - roof line projection.
   - roof pitch.
   - elevation symbols.
   - elevation material selection.
   - rules for elevation dimensions.
   - the effects of landscaping on the appearance of a structure.
4. In a check for competence, students should be able to perform these skills:
   - project and draw all four elevations from a floor plan, including dimensions.
   - make changes in elevation designs.
   - draw the elevation symbols for windows, doors, and building materials.
   - draw landscaping effects on an elevation.

ANSWERS TO END-OF-CHAPTER EXERCISES

Many end-of-chapter exercises ask students to perform an activity, such as draw a floor plan or photograph local structures. As a result, no "answer" is required and none is given here.

1–3. No answers required.

4. Answers will vary.

5–7. No answers required.

Date	M	Tu	W	Th	F

Lesson Plan for Chapter 17 (Text pages 281–307)

Sectional, Detail, and Cabinetry Drawings

ESSENTIAL ELEMENTS

For a complete listing of the essential elements, see the "Program Development" section in this guide. Essential elements for the course: 3.1, 3.5, 5.3, 6.5, 6.6, 6.7, 6.9, 6.12, 6.15

CHAPTER 17 OUTLINE

Sectional Drawings
- Design Concept Sections
- Presentation Sections

Full Sections
- The Cutting Plane
- Symbols
- Scale
- Sectional Dimensions
- Steps in Drawing Full Sections

Detail Sections
- Vertical Wall Sections
- Horizontal Wall Sections
- Window Sections
- Door Sections

Cabinetry and Built-in Component Drawings
- Cabinet Construction and Types
- Built-in Components
- Dimensioning Cabinetry and Built-ins

OBJECTIVES

Following are the objectives that appear at the beginning of Chapter 17 in the textbook. Students are expected to learn to:

- describe types of sectional drawings.
- communicate views of sections based on a cutting plane.
- draw sections, using correct section symbols and proper dimensioning.
- evaluate when a detail sectional drawing is needed.
- read and prepare detail drawings.
- design and prepare cabinet drawings.

USING THE COLOR IMAGE MASTERS AND IMAGE MASTERS

The following color image masters (CIM), image masters (IM), and handout masters (HO) are appropriate for use with this chapter. Teaching suggestions are included.

CIM-17A: Entertainment Center

- Explain how quality cabinet design and construction is a specialized skill. Simple cabinets can be built in on-site. Drawings for these are included in floor plans and interior elevations. Added details are keyed to more fully dimensioned detail drawings. However, elaborate or complex cabinetry involving doors, drawers, lighting, and integrated electronics equipment, as shown in this picture, are manufactured off-site by a cabinet specialist. The overall outlines and dimensions are usually shown on the floor plan and interior elevation, but the detail drawings are prepared by the manufacturer.
- Ask students why they think a unit such as this would not be built on-site.

IM-17A: Comparing Types of Drawings

- Point out that the top drawing shows a cutting plane line over a pictorial of a post and beam. In the center drawing this is done orthographically, and the bottom drawings show elevation sections.

- Ask students to explain the difference between section A-A and B-B.

IM-17B: Molding Details

- Use to show how sections for molding details are drawn.
- Mark other areas of the drawing and have students make sections; have them sketch a pictorial of this section.
- Ask students to identify the different types of moldings shown.

IM-17C: Pool Detail

- Use to show the need for more dimensioning detail than can be included on a complete site plan. This detail shows the same outlines as on the site plan; however, it also gives the installer all the dimensions needed to build the pool.
- For extra credit, have students create an additional elevation detail to describe the shape created by the various depth profiles.

HO-17A: Section Symbols for Common Building Materials

- Have students keep this handout in their notebooks for use when creating sections.

HO-17B: Elevation Symbols for Common Building Materials

- Have students keep this handout in their notebooks for use when creating elevations.

IM-17D Exterior Wall Detail

- This drawing shows a typical section of an exterior footing, floor, and wall intersection.

IM-17E Stair System Details

- This illustration includes a comparison of a plan, elevation, and pictorial view of a stair system.

IM-17F Stair and Header Intersection

- This drawing shows a section through a stair and header intersection.

EVALUATION

1. Assign the exercises at the end of the textbook chapter. When applicable, check students' answers against the answers given in the following section.
2. Test student mastery of factual material by using Chapter 17 Review found in the "Chapter Reviews" section of this binder.
3. In a check for competence, students should demonstrate understanding of:
   - the concepts of a section and of a cutting plane.
   - how to recognize material sections.
   - the concepts of section details and of vertical and horizontal sections.
   - the concept of sectional cabinet detail drawings.
4. In a check for competence, students should be able to perform these skills:
   - draw both longitudinal and transverse sections of a house.
   - draw section details of a wall, eave, footer, window, door, and kitchen floor cabinet.

ANSWERS TO END-OF-CHAPTER EXERCISES

Many end-of-chapter exercises ask students to perform an activity, such as draw a floor plan or photograph local structures. As a result, no "answer" is required and none is given here.

1–5. No answers required.

6. Custom made and modular; quality depends upon materials, joints, hardware, finish, and accuracy of construction and installation.
7. Hardwood, softwood, stranded lumber, laminates, ceramic tile, marble, and synthetic materials.

8–9. No answers required.

Date	M	Tu	W	Th	F

Lesson Plan for Chapter 18 (Text pages 308–344)

Site Development Plans

ESSENTIAL ELEMENTS

For a complete listing of the essential elements, see the "Program Development" section in this guide. Essential elements for the course: 2.4, 3.5, 5.2, 5.3, 6.2, 6.5, 6.6, 6.7, 6.9, 6.11

CHAPTER 18 OUTLINE

- Site Analysis
  - Suitability Levels
- Zoning Ordinances
  - Structural Types
  - Land Coverage and Setbacks
  - Density Zoning
  - Building Permits
- Survey Plans
  - Establishing Dimensions
  - Symbols on Survey Drawings
  - Guidelines for Drawing Surveys
  - Geographical Survey Maps
  - Plats
- Plot Plans
  - Guidelines for Drawing Plot Plans
  - Variations
- Landscape Plans
  - Guidelines for Drawing Landscape Plans
  - Phasing
- Landscape Rendering
  - Rendering Media
  - Plan Rendering
  - Elevation Rendering
- Site Details and Schedules

OBJECTIVES

Following are the objectives that appear at the beginning of Chapter 18 in the textbook. Students are expected to learn to:

- identify the major elements used in site design.
- understand the role and uses of zoning ordinances in the design process.
- draw survey, plat, and plot plans.
- understand the polar coordinate system and its application to site plans.
- design, draw, and render landscape plans and elevations.

USING THE COLOR IMAGE MASTERS AND IMAGE MASTERS

The following color image masters (CIM), image masters (IM), and handout masters (HO) are appropriate for use with this chapter. Teaching suggestions are included.

CIM-18A: Rendering Trees

- This illustration shows the steps used to render trees on landscape plans. Point out the noon shadowing in step 3. Ask students to draw the shadows that would appear in early morning or late afternoon views.

IM-18A: Terrain Profile

- Use to show the relationship between a terrain profile and contour interval spacing.
- Ask students to point out and define terminology used to describe the different areas of the profile.

IM-18B: Site Profile

- Use to show the intersection between contour interval lines and a cutting plane line. A right-angle projection is made from where the cutting plane line and contour lines meet to each corresponding strata line. When the points are connected, a profile is created.

IM-18C: Hills and Depressions

- Use to show how dimensions are indicated when the depth of a depression or the height of a hill falls between contour intervals.
- Give students dimensions with which to work and have them sketch similar elevations and depressions.

HO-18A: Setback Requirements

- The requirements shown are typical for areas zoned for single-family dwellings (R-15). Point out differences among the front, rear, and side setbacks and the resulting envelope (buildable area).
- Have students make a similar sketch of a property in their area.

HO-18B: Development Plat

- This illustration includes a key map to the plat. Have students refer to this plat for details when preparing their own plat plan.

HO-18C: Determining Bearings

- Use this to show the basic steps in determining the bearings needed to describe a property.
- Change the angle of one of the property lines and have students recalculate the new bearings in degrees, minutes, and seconds.

HO-18D: Site Plan Symbols

- Have students keep this handout in their notebooks for use when creating site plans.

HO-18E: Topographic Symbols for Large Tract Surveys and USGS Maps

- Have students keep this handout in their notebooks for use when studying surveys and maps.

IM-18D Site Profile

- This drawing includes a site plan with a projected elevation profile.

IM-18E Moisture Prevention

- Use this drawing to simplify and illustrate the common methods of altering site contours to reduce moisture accumulation.

HO 18F Plant List

- Due to limited space only a partial plant list is shown with the site plan in Fig. 18.39. This handout is the complete plant list which relates to the plan in Fig. 18.39 in the text.

EVALUATION

1. Assign the exercises at the end of the textbook chapter. When applicable, check students' answers against the answers given in the following section.
2. Test student mastery of factual material by using Chapter 18 Review found in the "Chapter Reviews" section of this binder.
3. In a check for competence, students should demonstrate understanding of:
   - how to orient a house on a lot, the relationship of a lot to its surroundings, and the relationship of site size and position to house orientation.
   - the effect of sun exposure, wind, and noise on a home.
   - active and passive solar orientation.
   - how to phase in landscaping through several steps.
   - the procedures for drawing a plot plan and the guide for a survey plan.
   - the basic use of a transit.
   - the background needed to be a surveyor, landscape architect, and a civil engineer.
4. In a check for competence, students should be able to perform these skills:
   - draw a plot plan using symbols and dimensions.
   - orient a house on a plot plan to take the best advantage of solar energy.
   - draw a landscape plan and landscape symbols.
   - draw a survey plan having dimensions, elevations, contour lines, and compass azimuths.
   - recognize local plants and trees.

ANSWERS TO END-OF-CHAPTER EXERCISES

Many end-of-chapter exercises ask students to perform an activity, such as draw a floor plan or photograph local structures. As a result, no "answer" is required and none is given here.

1. Environmental influences: slope, soils, vegetation, rare and endangered species, hydrology (surface water, flood hazard, ground water, wetlands), climate and microclimate, geology, wildlife and habitat, visual character, existing site, street layout and topography, existing land use and zoning, historical significance and preservation, available utilities.

   Human influences: external factors (noise and site accessibility), demographics, socio-economic forecasts.
2. Building codes limit building heights in order to allow neighbors maximum access to views, air circulation, and sunlight; this height is often 35 ft. The daylight plane ordinance limits the amount of space used in upper floors to allow more light to reach adjacent properties on the north side.
3. No answer required.
4. Answers will vary.
5. No answer required.

6–8. Answers will vary.

9–13. No answers required.

Date	M	Tu	W	Th	F

Lesson Plan for Chapter 19 (Text pages 345–360)

Pictorial Drawings

ESSENTIAL ELEMENTS

For a complete listing of the essential elements, see the "Program Development" section in this guide. Essential elements for the course: 3.3, 3.5, 5.3, 6.3, 6.4, 6.5, 6.6, 6.7

CHAPTER 19 OUTLINE

Types of Pictorial Projection
Perspective Drawings
Projection of Exterior Perspective Drawings
- One-Point Perspective
- Steps in Drawing a One-Point Perspective
- Two-Point Perspective
- Three-Point Perspective
- Projection Devices

Projection of Interior Perspective Drawings
- Isometric Drawings
- One-Point Perspective
- Two-Point Perspective

OBJECTIVES

Following are the objectives that appear at the beginning of Chapter 19 in the textbook. Students are expected to learn:

- to differentiate between isometric, oblique, and perspective drawings.
- geometric principles involved in projecting lines (from a given point or at a constant angle) to create 3-D images.
- to apply the principles of perspective drawing to create interior and exterior pictorial drawings.
- projection methods for drawing all types of pictorials.

USING THE COLOR IMAGE MASTERS AND IMAGE MASTERS

The following color image masters (CIM), image masters (IM), and handout masters (HO) are appropriate for use with this chapter. Teaching suggestions are included.

CIM-19A: Computer-Generated Perspective Drawing

- This illustration shows two phases of a computer-generated drawing. In the first view, different colors indicate different layers. Ask students to point out hidden lines.
- In the second view, hidden lines have been removed and surface treatments added. Ask students the purpose of adding the car and landscape features.

IM-19A: Projection Lines

- Use to show the lines connecting vanishing points.
- Ask students to identify which side of the drawing is done with one-point perspective and which side with two-point perspective.

IM-19B Floor Plan Pictorial

- This drawing shows a floor plan pictorial compared to an exterior pictorial.

IM-19C Two Point Perspective

- Shown here is a comparison of an isometric and a perspective view.

IM-19D Section and Perspective

- To describe the difference between an elevation and a pictorial drawing this illustration compares an interior elevation with a perspective of the same area.

IM-19E Interior Perspective

- This drawing shows the application of a one-point perspective on an interior view.

EVALUATION

1. Assign the exercises at the end of the textbook chapter. When applicable, check students' answers against the answers given in the following section.
2. Test student mastery of factual material by using Chapter 19 Review found in the "Chapter Reviews" section of this binder.
3. In a check for competence, students should demonstrate understanding of:
   - the theories of one-point, two-point, and three-point exterior perspective.
   - methods of perspective estimation and projection.
   - theories of one-point and two-point interior perspective.
   - the relationship of the viewer to the horizon, vanishing points, and finished perspective drawing.
4. In a check for competence, students should be able to perform these skills:
   - draw both a one-point and two-point exterior perspective of a house.
   - draw both a one-point and two-point interior perspective.
   - draw an isometric floor plan.

ANSWERS TO END-OF-CHAPTER EXERCISES

Many end-of-chapter exercises ask students to perform an activity, such as draw a floor plan or photograph local structures. As a result, no "answer" is required and none is given here.

1–9. No answers required.

Date	M	Tu	W	Th	F

Lesson Plan for Chapter 20 (Text pages 361–372)
Architectural Renderings

ESSENTIAL ELEMENTS

For a complete listing of the essential elements, see the "Program Development" section in this guide. Essential elements for the course: 3.4, 3.5, 4.1, 4.2, 5.3, 6.2, 6.5, 6.7

CHAPTER 20 OUTLINE

- Choosing Media for Rendering
    - Pencil Renderings
    - Ink Renderings
    - Watercolor and Wash Drawings
    - Oil and Acrylic Renderings
    - Felt Marker Renderings
    - Pressure-Sensitive Overlays
    - Media Combinations
- Showing the Effects of Light
    - Light Source and Shade
    - Shadow
- Texture
- Entourage
- Landscape
- Steps in Preparing a Rendering

OBJECTIVES

Following are the objectives that appear at the beginning of Chapter 20 in the textbook. Students are expected to learn to:

- recognize the wide selection of media available for renderings.
- evaluate when to use which media to achieve an artistic effect.
- add realism to drawings by the use of shading, shadows, texture, entourage, and landscapes.
- follow the correct sequence for preparing a rendering.

USING THE COLOR IMAGE MASTERS AND IMAGE MASTERS

The following color image masters (CIM), image masters (IM), and handout masters (HO) are appropriate for use with this chapter. Teaching suggestions are included.

CIM-20A: Ink and Wash Drawing

- Point out to students that line work was completed first, then watercolors were added, first to the building, then to foliage and sky. Ask why it was advantageous to use a very light color on the building.

CIM-20B: Pen and Ink Rendering

- This pen and ink rendering is really an elevation drawing. Ask students to identify details that make the illustration look like a pictorial, although no angles were used to indicate perspective. [Use of shadows and textures, light foreground, darker background.]

CIM-20C: Watercolor Rendering

- Watercolors were the predominant medium used to prepare this rendering. Ask students to look closely at the line work used for outlining, shadows, and textures. Point out the use of wash techniques in adding reflections to the windows and density to the foliage, which adds depth.
- Ask students how the roof line is emphasized.

HO-20A: Rendering Process

- Often, students add finishing details too soon to a rendering. Give this handout to them to help them follow the most efficient sequence when developing a combination ink and pencil rendering.

EVALUATION

1. Assign the exercises at the end of the textbook chapter. When applicable, check students' answers against the answers given in the following section.
2. Test student mastery of factual material by using Chapter 20 Review found in the "Chapter Reviews" section of this binder.
3. In a check for competence, students should demonstrate understanding of:
   - the types of pencils used for drawing textures.
   - the theories of texture and shadow.
   - rendering media.
4. In a check for competence, students should be able to perform these skills:
   - create texture and pattern with a soft pencil.
   - draw interior and exterior perspective, then render and shade walls, landscaping, windows, doors, fences, people, and autos.

ANSWERS TO END-OF-CHAPTER EXERCISES

Many end-of-chapter exercises ask students to perform an activity, such as draw a floor plan or photograph local structures. As a result, no "answer" is required and none is given here.

1–5. No answers required.

Date	M	Tu	W	Th	F

Lesson Plan for Chapter 21 (Text pages 373–382)

Architectural Models

ESSENTIAL ELEMENTS

For a complete listing of the essential elements, see the "Program Development" section in this guide. Essential elements for the course: 3.5, 4.1, 4.2, 5.3, 6.4, 6.7, 6.13

CHAPTER 21 OUTLINE

Design Study Models
- Solid Form Models
- Basic Layout Models
- Structural Models
- Interior Design Models

Presentation Models
- Contour-Interval Models
- Detailed Models

Steps in Constructing a Model

Computer Modeling

OBJECTIVES

Following are the objectives that appear at the beginning of Chapter 21 in the textbook. Students are expected to learn to:

- describe architectural models made for design study purposes.
- explain the differences between presentation and design study models.
- tell what input is needed to create a computer model.
- construct an architectural model.

USING THE COLOR IMAGE MASTERS AND IMAGE MASTERS

The following color image masters (CIM), image masters (IM), and handout masters (HO) are appropriate for use with this chapter. Teaching suggestions are included.

CIM-19A: Computer-Generated Perspective Drawing

- Re-use this color transparency from Chapter 19 to show an example of a model created on a computer.

EVALUATION

1. Assign the exercises at the end of the textbook chapter. When applicable, check students' answers against the answers given in the following section.
2. Test student mastery of factual material by using Chapter 21 Review found in the "Chapter Reviews" section of this binder.
3. In a check for competence, students should demonstrate understanding of:
   - the advantages of using three-dimensional structural models.
   - building procedures for a structural model.
   - model costs.
   - types of materials and tools used for model building.

4. In a check for competence, students should be able to perform these skills:
   - build a structural model.
   - complete the interior of a structural model.

ANSWERS TO END-OF-CHAPTER EXERCISES

Many end-of-chapter exercises ask students to perform an activity, such as draw a floor plan or photograph local structures. As a result, no "answer" is required and none is given here.

1. To check basic design ideas and for sales purposes.
2. Solid form models are used to check the overall proportions of a building; basic layout models are used to check overall layout and function of a design; structural models are used to check unique structural methods or to study framing options; interior design models are used to show individual room designs.
3. Details of the building's actual appearance and the building site.
4. (1) Create a floor plan base; (2) create the outer walls for every elevation; (3) create interior partitions; (4) attach window trim and glazing and create doors; (5) glue exterior walls to the floor plan base and to each other, then glue interior partitions in place; (6) finish wall surfaces and apply any wall or floor coverings; (7) install cabinetry and fixtures; (8) construct the roof and chimney; (9) add outdoor areas and property features; (10) add landscape features.

5–7. No answers required.

Date	M	Tu	W	Th	F

Lesson Plan for Chapter 22 (Text pages 383–397)

Principles of Construction

ESSENTIAL ELEMENTS

For a complete listing of the essential elements, see the "Program Development" section in this guide. Essential elements for the course: 3.4, 3.5, 5.1, 5.3, 6.7, 6.8, 6.9, 6.12, 6.13

CHAPTER 22 OUTLINE

Structural Design
- Structural Forces
- Strength of Materials

Modular Construction
- Modular Components
- Manufactured Buildings
- Modular Drawings

OBJECTIVES

Following are the objectives that appear at the beginning of Chapter 22 in the textbook. Students are expected to learn to:

- name and define the physical forces that act on a building.
- describe the factors that determine the strength of structural components.
- draw a modular floor plan, elevation, and detail drawing.

USING THE COLOR IMAGE MASTERS AND IMAGE MASTERS

The following color image masters (CIM), image masters (IM), and handout masters (HO) are appropriate for use with this chapter. Teaching suggestions are included.

IM-22A: Forces and Loads

- Remind students of the effects of strength, load, and length of span on the ability of building members to resist forces and stresses. Material, size, shape, placement, and spacing must be carefully planned. If any of these factors change during building, another element may need to be changed to compensate.

IM-22B: Beam Under Stress

- Use to show how a load creates compressive stress on the top part of a member while also creating tension on the bottom. Reinforcing the bottom half balances the stress. Larger members can obviously support greater loads than smaller members of the same material.
- Ask students what will happen if the load is not balanced.

IM-22C: Modular Door Unit

- Emphasize that accurate dimensions for rough openings are critical to the installation of modular doors. Dimensions must be clearly illustrated and labeled on the framing drawing.
- Have students sketch a wall section with a rough opening for a modular door or window, giving dimensions.

IM-22D: Notches

- Point out that builders must avoid excess notching of support members, especially horizontal members. Notching reduces the structural size of the member by the depth of the notch. Explain how blocking can compensate, but blocking must be tight, or it is of little value.

EVALUATION

1. Assign the exercises at the end of the textbook chapter. When applicable, check students' answers against the answers given in the following section.
2. Test student mastery of factual material by using Chapter 22 Review found in the "Chapter Reviews" section of this guide.
3. In a check for competence, students should demonstrate understanding of:
   - the basic principles of wood frame construction, heavy timber construction, masonry and concrete systems, and steel construction.
4. In a check for competence, students should be able to perform this skill:
   - draw construction details for various building materials.

ANSWERS TO END-OF-CHAPTER EXERCISES

Many end-of-chapter exercises ask students to perform an activity, such as draw a floor plan or photograph local structures. As a result, no "answer" is required and none is given here.

1. Compression forces push on objects; tension forces pull on objects; shear forces tend to make one part of an object slide past another part; torsion forces twist an object. Examples of how forces can be resisted will vary.
2. A live load is the weight of all movable objects, such as people and furniture, as well as the weight of snow and the force of the wind. Dead loads are the weight of building materials and permanently installed components. The total of both makes up the building load. These loads must be accounted for in terms of building design, materials used, and the size and spacing of members.
3. Answers will vary.
4. Different structural materials vary in their ability to resist stress and support loads. Larger members can support greater loads than smaller members. Different shapes can add to load-bearing capacity. Members with their widest dimension parallel to the direction of load will resist heavier loads than those placed with their smallest dimension parallel to the direction of load. Spacing determines how many members share the load.
5. Answers will vary.

6–9. No answers required.

Date	M	Tu	W	Th	F

Lesson Plan for Chapter 23 (Text pages 398–426)

Foundations and Fireplace Structures

ESSENTIAL ELEMENTS

For a complete listing of the essential elements, see the "Program Development" section in this guide. Essential elements for the course: 1.2, 3.1, 3.5, 5.2, 5.3, 6.5, 6.6, 6.7, 6.8, 6.11, 6.12

CHAPTER 23 OUTLINE

Foundation Materials and Components
- Bearing Surface
- Concrete
- Concrete Block
- Reinforcing Bars
- Wire Mesh
- Footings

Types of Foundations
- T-Foundations
- Slab Foundations
- Pier-and-Column Foundations
- Permanent Wood Foundations

Fireplaces
- Fuel Types
- Types of Fireplace Openings
- Fireplace Construction
- Fireplace Detail Drawings

Foundation Drawings
- Layout
- Excavations and Forming
- Foundation Plans
- Detail Drawings

OBJECTIVES

Following are the objectives that appear at the beginning of Chapter 23 in the textbook. Students are expected to learn to:

- describe the types of foundations.
- identify the components and materials used in foundations.
- design a fireplace with sufficient structural support and appropriate safety components.
- draw foundation plans.
- relate the layout and excavations for a building to the type of foundation it will have.

USING THE COLOR IMAGE MASTERS AND IMAGE MASTERS

The following color image masters (CIM), image masters (IM), and handout masters (HO) are appropriate for use with this chapter. Teaching suggestions are included.

IM-23A: Relationship Between Section and Pictorial Drawings

- Use this to show the nomenclature and differences between a section drawing and a pictorial drawing of the same subject.
- Ask students to point out useful features of each type of drawing.

IM-23B: Elevation and Pictorial

- Use to show an elevation section illustrating the left side of the pictorial section.
- Have students sketch an elevation section illustrating the right side (90° from the elevation shown).

IM-23C: Fireplace Foundation

- Use to show the subhearth structure and footing needed to support the weight of the fireplace and to tie the hearth into the floor structure.

- Ask students how a cantilever is used in this design.

IM-23D: **Foundation Detail**

- Use to show the configuration of different widths of concrete block forming a pilaster and encasement for a wood beam.
- Have students sketch an elevation section and plan view of this detail.

IM-23E: **Foundation Insulation**

- Use to show the application of batt insulation to the perimeter of a T-foundation.
- Have students sketch an elevation section of this detail.

IM-23F: **T-foundation and Slab Foundation**

- This illustration shows an elevation section, plan view, and pictorial section of a slab foundation on the left and a pictorial section of a T-foundation on the right. Have students sketch a plan and elevation section of the T-foundation similar to what has been done for the slab.

HO-23A: **Foundation Chart**

- Remind students that the amount of concrete needed for a slab foundation is found by calculating cubic feet (length x width x depth). Cubic feet are then converted to cubic yards because this is how concrete is sold. (There are 27 cubic feet in one cubic yard.)
- Suggest that students keep this chart in a notebook for easy reference when calculating for slab thicknesses of 4″, 5″, and 6″.

IM-23G Footing Section

- One of the most common structural detail drawings is a section through a typical concrete footer and slab intersection as shown here.

IM-23H Fireplace Construction

- This drawing illustrates a typical fireplace construction with a cut-away pictorial drawing.

EVALUATION

1. Assign the exercises at the end of the textbook chapter. When applicable, check students' answers against the answers given in the following section.
2. Test student mastery of factual material by using Chapter 23 Review found in the "Chapter Reviews" section of this guide.
3. In a check for competence, students should demonstrate understanding of:
   - the principles of slab, column, and T-foundation design and construction.
   - how to read girder and joist span tables.
   - the terminology for foundation parts.
   - different types of fireplaces, the foundations required, the names for different parts, and how a fireplace works.
4. In a check for competence, students should be able to perform these skills:
   - design and draw a slab foundation and section details.
   - design and draw a T-foundation, including floor system and section details.
   - calculate yards of concrete for foundation work.
   - calculate yards of earth to move for foundation forms.
   - create a working drawing and sections of a fireplace.
   - draw different types of fireplaces on a floor plan.

ANSWERS TO END-OF-CHAPTER EXERCISES

Many end-of-chapter exercises ask students to perform an activity, such as draw a floor plan or photograph local structures. As a result, no "answer" is required and none is given here.

1–9. No answers required.

Date	M	Tu	W	Th	F

Lesson Plan for Chapter 24 (Text pages 427–445)

Wood-Frame Systems

ESSENTIAL ELEMENTS

For a complete listing of the essential elements, see the "Program Development" section in this guide. Essential elements for the course: 3.4, 3.5, 4.2, 5.1, 5.3, 6.8, 6.9, 6.12

CHAPTER 24 OUTLINE

Skeleton-Frame Construction
- Materials
- Wood-Framing Methods
- Board-Foot Measure

Post-and-Beam Construction
- Floor Construction
- Wall Construction
- Roof Construction
- Structural Timber Members

OBJECTIVES

Following are the objectives that appear at the beginning of Chapter 24 in the textbook. Students are expected to learn to:

- differentiate between skeleton-frame and post-and-beam construction.
- identify major characteristics of lumber, plywood, and structural timber.
- calculate the number of board feet in a piece of lumber.

USING THE COLOR IMAGE MASTERS AND IMAGE MASTERS

The following color image masters (CIM), image masters (IM), and handout masters (HO) are appropriate for use with this chapter. Teaching suggestions are included.

IM-24A: Heavy Timber Intersections

- This illustration shows views of common heavy timber intersections as drawn on plan and elevation framing drawings. Pictorials are shown at the left. Cover the plan and elevation views and have the students sketch those using the pictorial drawings as a reference.
- Uncover the plan and elevation views and have students check their work and make any corrections.

IM-24B: Gusset Plate and Joist Hanger

- This steel gusset plate and joist hanger attach a joist and post to a beam. Have students sketch a section drawing of this detail as seen from the right or left sides.

IM-24C: Split Ring Connector

- Have students sketch side and top views of the exploded view on the bottom right.

IM-24D Building Units

- Use this illustration to compare US customary modular building units with SI Metric units.

IM-24F Residential Framing

- Framing methods are shown with plan, elevation, or pictorial sections. This drawing shows typical residential skeleton framing methods with a pictorial wall section.

EVALUATION

1. Assign the exercises at the end of the textbook chapter. When applicable, check students' answers against the answers given in the following section.

2. Test student mastery of factual material by using Chapter 24 Review found in the "Chapter Reviews" section of this binder.
3. In a check for competence, students should demonstrate understanding of:
   - the concepts, methods used, and types of wood construction.
4. In a check for competence, students should be able to perform this skill:
   - sketch wood-frame details.

ANSWERS TO END-OF-CHAPTER EXERCISES

Many end-of-chapter exercises ask students to perform an activity, such as draw a floor plan or photograph local structures. As a result, no "answer" is required and none is given here.

1. In skeleton-frame construction, small structural members are joined in such a way that they share the loads of the structure; and when members are covered, they form walls, floors, and roof. Post-and-beam construction uses larger structural members, and they are spaced farther apart; some of the columns, beams, and planks may remain exposed.
2. In platform framing, the second floor is constructed directly on first-floor exterior walls. In balloon-framed buildings, the first-floor joists rest directly on a sill plate, and the second floor joists bear on ribbon strips set into continuous two-level studs.
3. Wood I beams, truss joists, stressed-skin or sandwich panels, foam sandwich panels, strand lumber, and recycled thermo-plastics.
4. Answers will vary.

5–6. No answers required.

7. Yard lumber is used for most light framing members, such as bracing, sheathing, subfloors, and casings. Structural lumber is used for load-bearing members. Factory, or shop, lumber consists of light members which are finished at a mill and used for trim, molding, and door and window sashes.
8. 4800 board feet ($2 \times 6 \times 12 \times 12 = 12 \times 400 = 4800$)
9. 1 1/2″ × 3 1/2″ (2 × 4)
3 9/16″ × 5 1/2″ (4 × 6)
10. 1/8″ to 1 1/4″
11. Hardwoods are used for surfaces that must withstand wear, such as floors and railings, and for items that require a fine natural finish, such as cabinets and furniture. Softwoods are used for structural members, such as joists, rafters, studs, sheathing, and formwork.

Date	M	Tu	W	Th	F

Lesson Plan for Chapter 25 (Text pages 446–460)

Masonry and Concrete Systems

ESSENTIAL ELEMENTS

For a complete listing of the essential elements, see the "Program Development" section in this guide. Essential elements for the course: 3.5, 4.2, 4.3, 5.3, 6.8, 6.9, 6.12

CHAPTER 25 OUTLINE

Masonry Construction Systems
- Masonry Materials
- Masonry Walls

Concrete Construction Systems
- Reinforced Concrete
- Prestressed Concrete
- Concrete Structural Members
- Cast-in-Place Concrete
- Precast Concrete
- Slab Component Systems
- Concrete Joints
- Concrete Wall Systems

OBJECTIVES

Following are the objectives that appear at the beginning of Chapter 25 in the textbook. Students are expected to learn to:

- identify the types of masonry materials used in construction.
- describe four types of masonry walls.
- describe ways to strengthen concrete and prevent deflection.
- explain how concrete is used for slabs and other structural components.

USING THE COLOR IMAGE MASTERS AND IMAGE MASTERS

The following color image masters (CIM), image masters (IM), and handout masters (HO) are appropriate for use with this chapter. Teaching suggestions are included.

IM-25A: Masonry Walls

- Have students sketch, dimension, and label member sizes for these exterior wall sections.

IM-25B: Roof Fascia

- This roof fascia detail includes information on the attachment of the top plate and the intersection of the fascia plate and decking. Have students sketch a section of this detail as viewed from the left side.

IM-25C Poured Concrete Wall

- This drawing shows a poured concrete wall section.

IM-25D Concrete Block Wall

- This drawing shows a concrete block section. Have students compare this drawing with the one in IM-25C.

EVALUATION

1. Assign the exercises at the end of the textbook chapter. When applicable, check students' answers against the answers given in the following section.
2. Test student mastery of factual material by using Chapter 25 Review found in the "Chapter Reviews" section of this guide.

3. In a check for competence, students should demonstrate understanding of:
   - concepts relating to methods of masonry construction.
4. In a check for competence, students should be able to perform this skill:
   - sketch masonry construction details.

ANSWERS TO END-OF-CHAPTER EXERCISES

Many end-of-chapter exercises ask students to perform an activity, such as draw a floor plan or photograph local structures. As a result, no "answer" is required and none is given here.

1. Common brick is generally used in unexposed areas; face brick is used in exposed areas that require accurate dimensions, absorption control, and consistent color and texture.
2. Solid masonry walls are solid all the way through and may have rebars added for strength; masonry cavity walls are hollow in the middle; masonry-faced walls may be of any type of masonry faced with another material, such as tile or brick, and bonded to form one wall structurally; masonry-veneer walls are two walls side by side, but they do not form a single unit and the veneer wall remains a decorative facade.
3. Reinforced concrete contains steel bars added for strength. Prestressed concrete may undergo pretensioning, in which deformed steel bars are stretched between anchors and the concrete poured around them; this creates a continual state of compressive stress which counteracts load stresses. Or it may undergo post-tensioning, in which steel bars are placed inside tubes embedded in the concrete and then stretched, which also counteracts load stresses.

   Concrete structural members, such as slabs, columns, beams, girders, and lintels may be all reinforced with high-tensile-strength steel.

   Cast-in-place concrete is poured at the site and may contain steel reinforcements.

   Precast concrete is poured into wood, metal, or plastic molds and is usually reinforced; after hardening, it is placed in position.

   In one-way slab component systems, all the rebars are laid parallel, and the girders, which rest on columns, are parallel to the rebars. In a two-way slab system, the rebars, girders, and beams are perpendicular to one another.

   Conventional concrete wall systems are poured between wood or metal forms and may be reinforced; insulated concrete walls are poured between polystyrene panels, which become part of the wall; other systems use interlocking hollow-core blocks filled with concrete.

4–5. No answers required.

Date	M	Tu	W	Th	F

Lesson Plan for Chapter 26 (Text pages 461–474)

Steel and Reinforced-Concrete Systems

ESSENTIAL ELEMENTS

For a complete listing of the essential elements, see the "Program Development" section in this guide. Essential elements for the course: 3.3, 3.5, 4.2, 4.3, 5.3, 6.8, 6.9, 6.12

CHAPTER 26 OUTLINE

Steel Building Construction
Steel Structural Members
- Plates
- Bars
- Steel Pipe and Structural Tubing
- Other Structural Steel Shapes

Steel Fasteners and Intersections
- Brackets
- Rivets
- Bolts
- Welds

Structural Steel Drawing Conventions

OBJECTIVES

Following are the objectives that appear at the beginning of Chapter 26 in the textbook. Students are expected to learn to:

- describe three types of steel construction and explain the basic purpose of each.
- identify manufactured steel forms and their function as structural members.
- read and interpret steel symbols, notations, identification, and measurements for working drawings.
- relate the types of fasteners and intersections of steel members to construction methods.

USING THE COLOR IMAGE MASTERS AND IMAGE MASTERS

The following color image masters (CIM), image masters (IM), and handout masters (HO) are appropriate for use with this chapter. Teaching suggestions are included.

IM-26A: Steel Framing

- Use to show steel framing in high-rise buildings. Point out the relationship of increased size to spacing and cross-bracing.
- Have students speculate about the framing method used in the nearest or most visible high-rise in their area.

HO-26A: Concrete Floor System

- Point out that this shows an elevation section and pictorial interpretation of a reinforced concrete floor system having precast concrete joists.
- Have students draw a plan view.

HO-26B: Steel Joist Detail

- Point out how the steel joists intersect and are fastened to headers with metal clips.
- Have students sketch an elevation section as viewed from the right of the pictorial drawing.

HO-26C Steel Framing

- This handout provides a written description of the advantages and disadvantages of residential steel framing.

EVALUATION

1. Assign the exercises at the end of the textbook chapter. When applicable, check students' answers against the answers given in the following section.
2. Test student mastery of factual material by using Chapter 26 Review found in the "Chapter Reviews" section of this guide.
3. In a check for competence, students should demonstrate understanding of:
   - concepts behind steel construction methods.
   - concepts behind steel-reinforced concrete construction methods.
4. In a check for competence, students should be able to perform these skills:
   - sketch steel construction details.
   - sketch steel-reinforced concrete construction details.

ANSWERS TO END-OF-CHAPTER EXERCISES

Many end-of-chapter exercises ask students to perform an activity, such as draw a floor plan or photograph local structures. As a result, no "answer" is required and none is given here.

1. Design drawings are symbolic and show only the position of each structural member using a single line; notations for shape, size, and weight are included. Shop drawings are complete orthographic engineering drawings showing the exact size and shape of each member, including every cut, hole, and method of fastening. Erection drawings show the method and order of assembling each member.
2. Steel can span greater distances and support greater loads than other construction materials; however, its disadvantage is that it is very heavy and creates heavy dead loads.
3. Beams are horizontal members placed on or between girders; girders are horizontal members that extend between columns; columns are vertical members that rest on footings or piers; purlins are beams that connect roof trusses or rafters.
4. Plates, bars, pipes, tubing, L-shapes, channels, S-shapes, W-shapes, M-shapes.

5–7. No answers required.

8. Brackets, rivets, bolts, welds.
9. No answer required.

Date	M	Tu	W	Th	F

Lesson Plan for Chapter 27 (Text pages 475–482)

Disaster Prevention Design

ESSENTIAL ELEMENTS

For a complete listing of the essential elements, see the "Program Development" section in this guide. Essential elements for the course: 1.1, 3.2, 3.5, 5.3, 6.10

CHAPTER 27 OUTLINE

Preventing Wind Damage
Preventing Earthquake Damage
Preventing Gas Leakage
  Natural and Propane Gas
  Carbon Monoxide (CO)
  Radon
Fire Prevention or Control
Water and Air Purification
Other Safety Considerations

OBJECTIVES

Following are the objectives that appear at the beginning of Chapter 27 in the textbook. Students are expected to learn to:

- describe the measures that can be taken during construction to minimize potential damage from natural disasters.
- describe how to prevent gas leakage.
- name ways to provide fire protection for a structure and its residents.
- discuss methods for ensuring clean air and water in a building.

USING THE COLOR IMAGE MASTERS AND IMAGE MASTERS

The following color image masters (CIM), image masters (IM), and handout masters (HO) are appropriate for use with this chapter. Teaching suggestions are included.

HO-27A: Dangers of Carbon Monoxide

- Discuss with students the kinds of poor construction that can cause carbon monoxide to build up within a home.

IM-27A Strong Tie Details

- The detail drawings shown here illustrate several methods of strengthening intersections with *strong ties*.

HO-27B Earthquakes

- This illustration describes construction design considerations related to earthquake damage prevention.

EVALUATION

1. Assign the exercises at the end of the textbook chapter. When applicable, check students' answers against the answers given in the following section.
2. Test student mastery of factual material by using Chapter 27 Review found in the "Chapter Reviews" section of this guide.

3. In a check for competence, students should demonstrate understanding of:
   - the various types of connectors used to strengthen structures.
4. In a check for competence, students should be able to perform these skills:
   - draw details of wood and metal connectors.

ANSWERS TO END-OF-CHAPTER EXERCISES

Many end-of-chapter exercises ask students to perform an activity, such as draw a floor plan or photograph local structures. As a result, no "answer" is required and none is given here.

1. Answers will vary.

2–3. No answers required.

4. Answers will vary.
5. No answer required.

Date	M	Tu	W	Th	F

Lesson Plan for Chapter 28 (Text pages 483–507)

Floor Framing Drawings

ESSENTIAL ELEMENTS

For a complete listing of the essential elements, see the "Program Development" section in this guide. Essential elements for the course: 3.5, 5.3, 6.4, 6.5, 6.6, 6.7, 6.9, 6.10, 6.12, 6.15

CHAPTER 28 OUTLINE

Types of Platform Floor Systems
Floor Framing Members
  Decking
  Joists
  Girders and Beams
Floor Framing Plans
  Floor Framing Plans for Wood
  Floor Framing Plans for Steel

OBJECTIVES

Following are the objectives that appear at the beginning of Chapter 28 in the textbook. Students are expected to learn to:

- identify the components of floor systems.
- draw a floor framing plan that shows all structural parts.
- draw details of sills, supports, and stairwells.

USING THE COLOR IMAGE MASTERS AND IMAGE MASTERS

The following color image masters (CIM), image masters (IM), and handout masters (HO) are appropriate for use with this chapter. Teaching suggestions are included.

IM-28A: Floor Deck System

- Use to show a pictorial section of a deck system for floor tile.
- Have students sketch a section detail as viewed from the left side of this drawing.

HO-28A: Rough Openings

- Use to show how to calculate the rough openings for common doors and windows.
- Have students calculate a rough opening for a 12-foot garage door or any other standard width not shown on this chart.

HO-28B: Spacing of Studs

- The spacing shown is typical in light construction. Point out the key showing the symbols used to identify different types of studs on a drawing.
- Have students use these symbols for reference in preparing stud layouts or details.

HO-28C: Floor System

- Use this to show the relationship between plan and elevation views of a floor system. The pictorial shows how the system might look to the eye.
- Have students sketch a section as viewed from the left side of the pictorial drawing.

HO-28D Floor Joist Direction Change

- This drawing illustrates one method of intersecting floor joists at right angles.

EVALUATION

1. Assign the exercises at the end of the textbook chapter. When applicable, check students' answers against the answers given in the following section.
2. Test student mastery of factual material by using Chapter 28 Review found in the "Chapter Reviews" section of this guide.

3. In a check for competence, students should demonstrate understanding of:
   - the different types of stud layouts and the symbols used.
   - rough and dressed lumber sizes.
   - the terminology for floor framing parts.
   - how to use floor framing tables.
   - the principles governing strength of materials and structural spans and spacing.
4. In a check for competence, students should be able to perform these skills:
   - draw a floor framing plan, listing all structural parts from tables, including sizes, spans, and spacing.
   - draw different corner-post layouts.
   - draw different partition intersections.

ANSWERS TO END-OF-CHAPTER EXERCISES

Many end-of-chapter exercises ask students to perform an activity, such as draw a floor plan or photograph local structures. As a result, no "answer" is required and none is given here.

1–6. No answers required.

Date	M	Tu	W	Th	F

Lesson Plan for Chapter 29 (Text pages 508–538)

Wall Framing Drawings

ESSENTIAL ELEMENTS

For a complete listing of the essential elements, see the "Program Development" section in this guide. Essential elements for the course: 3.5, 5.2, 5.3, 6.4, 6.5, 6.6, 6.7, 6.9, 6.10, 6.12, 6.15

CHAPTER 29 OUTLINE

Exterior Walls
- Framing Elevations
- Detail Drawings
- Window Framing Drawings
- Door Framing Drawings

Interior Walls
- Framing Elevations
- Details

Stud Layouts
- Dimensions
- Stud Details
- Modular Plans
- Steel Studs

OBJECTIVES

Following are the objectives that appear at the beginning of Chapter 29 in the textbook. Students are expected to learn to:

- draw an exterior wall framing elevation and plan.
- draw an interior wall framing elevation and plan.
- draw details and sections of walls.
- draw wall intersections.

USING THE COLOR IMAGE MASTERS AND IMAGE MASTERS

The following color image masters (CIM), image masters (IM), and handout masters (HO) are appropriate for use with this chapter. Teaching suggestions are included.

CIM-29A: **Walls and Second-Story Floor System**

- This photo shows framing for the southwest and southeast elevations shown in Fig. 14-20 in the student text and in the set of plans accompanying this instructor's guide. Have students locate the windows and corners on both levels in the floor plans and elevations.
- Ask students what the second floor overhang represents on the plans.

IM-29A: **Stud Layout and Framing Elevation**

- Point out how the interior elevation is based on the view indicated by the arrow on the pictorial drawing. The stud layout is the plan section of the framing wall elevation.
- Have students sketch another wall shown in the pictorial drawing and/or develop a related stud layout.

HO-29A: **Stud and Siding Wall**

- Use this to show a plan and elevation section of a wall having wood studs and horizontal siding.
- Ask students to use the pictorial to interpret the elevation and plan sections.

IM-29B: **Brick Wall Section**

- This drawing shows a full wall section of a brick veneer wall.

HO-29B Brick-Veneer Wood Framed Wall

- This drawing shows a sill elevation section and a related floor framing plan for a brick veneer wall.

EVALUATION

1. Assign the exercises at the end of the textbook chapter. When applicable, check students' answers against the answers given in the following section.
2. Test student mastery of factual material by using Chapter 29 Review found in the "Chapter Reviews" section of this guide.
3. In a check for competence, students should demonstrate understanding of:
   - the principles of framing, including wood and steel construction, loads, skeleton frames, post-and-beam construction, and prefabrication.
   - the advantages and disadvantages of each type of framing.
   - the different styles of steel frameworks.
   - the terminology for exterior wall framing parts.
   - the modular sizes for wall framing.
   - the variations among framing methods.
   - the different types of wall coverings.
   - optional methods for framing doors.
   - methods used to install drywall, paneling, and plaster.
4. In a check for competence, students should be able to perform these skills:
   - sketch examples of framing, including skeleton, post-and-beam, cantilever, and bearing wall framing.
   - make models showing skeleton, post-and-beam, cantilever, and bearing wall framing.
   - project from a floor plan and draw an exterior-wall framing elevation.
   - draw a panel framing plan for skeleton, post-and-beam, cantilever, and bearing wall framing.
   - draw wall details and sections.
   - draw an interior-wall framing plan.
   - draw details and sections of different types of wall intersections.

ANSWERS TO END-OF-CHAPTER EXERCISES

Many end-of-chapter exercises ask students to perform an activity, such as draw a floor plan or photograph local structures. As a result, no "answer" is required and none is given here.

1–9. No answers required.

Date	M	Tu	W	Th	F

Lesson Plan for Chapter 30 (Text pages 539–568)
Roof Framing Drawings

ESSENTIAL ELEMENTS

For a complete listing of the essential elements, see the "Program Development" section in this guide. Essential elements for the course: 3.5, 5.3, 6.4, 6.5, 6.6, 6.7, 6.10

CHAPTER 30 OUTLINE

- Roof Function
- Roof Framing Members
    - Wood Roof Members
    - Steel Roof Members
    - Concrete Roof Members
- Roof Framing Components
    - Trusses
    - Roof Panels
    - Cornices
    - Collar Beams and Knee Walls
    - Dormers
    - Chimney Details
- Roof Framing Drawings
    - Plans
    - Elevations
    - Dimensions
- Roof Pitch
- Roof Framing Methods for Wood
- Roof Framing Types
    - Gable Roof Framing
    - Gambrel Roof Framing
    - Hip Roof Framing
    - Dutch Hip Roof Framing
    - Shed Roof Framing
    - Flat Roof Framing
- Steel and Concrete Roof Framing Methods
- Roof-covering Materials
    - Sheathing
    - Roll Roofing
    - Roof Shingles, Shakes, and Tiles
    - Built-up Roofs
- Roof Appendages
    - Gutters and Downspouts
    - Flashing
    - Roof Ventilation
    - Skylights

OBJECTIVES

Following are the objectives that appear at the beginning of Chapter 30 in the textbook. Students are expected to learn to:

- describe roof framing members, components, and methods.
- calculate roof pitch.
- draw a roof framing plan showing structural members, sizes, pitch, and spacing.
- draw framing details and elevations.

USING THE COLOR IMAGE MASTERS AND IMAGE MASTERS

The following color image masters (CIM), image masters (IM), and handout masters (HO) are appropriate for use with this chapter. Teaching suggestions are included.

CIM-30A: Roof Framing

- This photo shows framing for the southwest and southeast elevations shown in Fig. 14-20 in the student text and in the set of plans accompanying this teacher's resource binder. It was taken from the kitchen area facing the kitchen truss joist roof framing and through the truss framing of the garage roof. Have students identify this vantage point on the floor plan.

- Have students identify on the plans the roof areas shown in the picture.

IM-30A: Attaching Collar Beams to Roof Rafters

- Have students sketch or draw a gable and section view of this area.

IM-30B: Roof Rafter, Side View

- Point out the notations relating to truss type and position, which can be seen only from a section 90 degrees from this one.

HO-30A: Eave Detail

- Use to show a pictorial drawing of an eave detail with roof covering materials removed to expose the rafters, studs, and soffit framing.
- Have students sketch or draw an elevation section of this detail.

HO-30B: Roof Intersection

- Use to show an elevation and pictorial section of a steel joist and roof decking intersection with a concrete block and veneer wall.
- Have students sketch an elevation section by referring to the pictorial drawing.

IM-30C Roof Framing Plan

- This drawing shows a roof framing plan.

IM-30D Rafter Intersections

- This drawing shows how rafter intersections are drawn.

IM-30E Roof Pitches

- This drawing illustrates typical roof pitches.

IM-30F Gable Roof Framing

- Shown here are several variations of gable roof framing.

IM-30G Hip and Gable Framing

- This illustration includes the terminology used to describe roof framing members.

IM-30H Dormer Framing

- Pictorial framing drawings in this illustration show framing methods for shed and gable dormers.

IM-30I Mansard Framing

- This illustration shows methods of framing mansard roofs.

IM-30J Gambrel Framing

- Methods of Gambrel framing are shown here.

IM-30K Knee Wall Framing

- The section and pictorial drawing in this illustration shows the construction of a typical knee wall.

EVALUATION

1. Assign the exercises at the end of the textbook chapter. When applicable, check students' answers against the answers given in the following section.
2. Test student mastery of factual material by using Chapter 30 Review found in the "Chapter Reviews" section of this guide.
3. In a check for competence, students should demonstrate understanding of:
   - the different types of roof framing and the terminology for roof framing parts.
   - dead and live loads for a roof.
   - different roof pitches.
   - different roof overhangs, covering materials, and shingle patterns.
   - built-up roofing.
4. In a check for competence, students should be able to perform these skills:
   - draw a roof framing plan listing structural sizes, spans, and spacing.
   - calculate bending and resisting movements.
   - draw standard roof trusses, different cornice framing details, and a framed roof dormer.
   - plan and draw methods of sun screening.
   - plan and draw roof gutters and downspouts.
   - specify flashing on a drawing.

ANSWERS TO END-OF-CHAPTER EXERCISES

Many end-of-chapter exercises ask students to perform an activity, such as draw a floor plan or photograph local structures. As a result, no "answer" is required and none is given here.

1. Simplified, single-line plans show the general relationship and spacing of the structural members;

detailed, double-line plans show the exact construction of intersections and joints, as well as the thickness of each member.

2. No answer required.
3. 1/6
4. Top chord, bottom chord, and webs; trusses prevent sags and cracks in roofs, can span larger distances than conventional framing, and can be used with steel members, masonry walls, or wood-framed walls; eliminates attic or crawl space, more expensive, must order trusses in advance.

5–6. No answers required.

7. Answers will vary.
8. 400 sq. ft.
9. No answer required.

Date	M	Tu	W	Th	F

Lesson Plan for Chapter 31 (Text pages 569–594)

Electrical Design and Drawings

ESSENTIAL ELEMENTS

For a complete listing of the essential elements, see the "Program Development" section in this guide. Essential elements for the course: 2.1, 3.2, 3.4, 3.5, 5.2, 5.3, 6.5, 6.6, 6.7, 6.8, 6.9, 6.10, 6.11, 6.13, 6.15

CHAPTER 31 OUTLINE

Electrical Principles
- Power Distribution
- Electrical Measurements
- Service Entrance
- Service Distribution
- Branch Circuits
- Ground-Fault Circuit Interrupter (GFCI)
- Electrical Conductors
- Calculating Total System Requirements

Lighting Design
- Light Measurements
- Types of Lighting
- Light Distribution
- Reflection
- Structural Light Fixtures

Developing and Drawing Electrical Plans
- Fixture and Device Selection
- Switches
- Electrical Outlets and Receptacles

Electrical Working Drawings
- Electrical Symbols
- Room Wiring Drawings

Electronic Systems
- Electronic System Drawings
- The Complete Plan

OBJECTIVES

Following are the objectives that appear at the beginning of Chapter 31 in the textbook.

Students are expected to learn to:

- plan and draw electrical circuits for a house on a floor plan.
- plan and draw lighting for each room in a house.
- calculate electrical measurements for each circuit.
- draw electrical and electronic symbols.
- design and draw an electronic building control system.

USING THE COLOR IMAGE MASTERS AND IMAGE MASTERS

The following color image masters (CIM), image masters (IM), and handout masters (HO) are appropriate for use with this chapter. Teaching suggestions are included.

CIM-31A: Lighting

- Point out that this photo shows how indoor and outdoor lighting can be combined to produce a dramatic effect and also provide for nighttime activities.

CIM-31B: Ceiling Lights

- Point out that this shows an example of the use of ceiling and soffit lighting for both general and specific tasks. (This is the same house shown in Fig. 1-18 in the student text.)
- Ask students to list the advantages and disadvantages to the user of the lighting shown here.

IM-31A: Electrical Plan

- Have students trace the connections between switches and fixtures and find the second floor switches or fixtures that are connected to the first floor.

HO-31A: First Floor Electrical Plan

- This is the first floor electrical plan for the set of drawings that accompany this instructor's guide. Have students match the letters with the letters found on the other drawings.
- Have students trace the connections between switches and fixtures.

IM-31B Electrical Plan

- This is an example of a basic electrical plan showing the application of common symbols.

HO-31B Planning Circuits and Outlets

- This illustration provides a summary of the planning of electrical circuits and the location of outlets.

EVALUATION

1. Assign the exercises at the end of the textbook chapter. When applicable, check students' answers against the answers given in the following section.
2. Test student mastery of factual material by using Chapter 31 Review found in the "Chapter Reviews" section of this guide.
3. In a check for competence, students should demonstrate understanding of:
   - the terminology used in electrical planning.
   - load requirements for electrical equipment.
   - wire sizes and safe current as well as safety factors relating to electrical planning.
   - the formulas for calculating voltage, amperage, and wattage.
   - the electrical symbols used on a floor plan.
   - the types of lighting, how light is dispersed, and light measurements.
4. In a check for competence, students should be able to perform these skills:
   - plan and draw electrical circuits for a house floor plan using the proper symbols.
   - plan service entry size.
   - figure maximum load, fuse size, and wire size for each circuit.
   - design and draw a low-voltage control system.
   - plan lighting requirements for each room in a house.

ANSWERS TO END-OF-CHAPTER EXERCISES

Many end-of-chapter exercises ask students to perform an activity, such as draw a floor plan or photograph local structures. As a result, no "answer" is required and none is given here.

1. 168.3 amperes
2. 8.57 amperes
3. Lighting, small appliance, and individual
4. Answers will vary but should name places wherever shock is possible, such as near water or pipes.
5. Footcandle (or candelas, lux in metric)
6. General, specific, decorative
7. Direct, indirect, semi-direct, semi-indirect, diffused

8–9. Answers will vary.

10. No answer required.
11. Answers will vary.

12–13. Answers will vary.

14. 6 ft. apart (except in the kitchen where they are 4 ft. apart) and 12–18 in. from the floor.

15–18. No answers required.

Date	M	Tu	W	Th	F

Lesson Plan for Chapter 32 (Text pages 595–616)

Comfort-Control Systems (HVAC)

ESSENTIAL ELEMENTS

For a complete listing of the essential elements, see the "Program Development" section in this guide. Essential elements for the course: 3.1, 3.5, 4.1, 4.2, 5.1, 5.2, 5.3, 6.5, 6.6, 6.7, 6.9, 6.10, 6.15

CHAPTER 32 OUTLINE

HVAC Plans and Conventions
Principles of Heat Transfer
- Methods of Heat Transfer
- Heat Measurement
- Insulation
- Heat Loss Calculations

Heating Systems
- Wood Plenum System
- Warm-Air Units
- Hydronic (Hot-Water) Units
- Steam Units
- Electric Heat

Cooling Systems
Heat Pumps
Ventilation
Air Filtration
HVAC Control Devices
Humidity Control
Passive Solar Systems
- Greenhouse Effect
- Rising Warm Air
- Passive Solar Methods

Active Solar Systems
- Collection
- Storage
- Distribution and Control
- Types of Active Solar Systems

OBJECTIVES

Following are the objectives that appear at the beginning of Chapter 32 in the textbook. Students are expected to learn to:

- plan and draw a mechanical heating and cooling system on a floor plan.
- use appropriate symbols to draw devices, ductwork, or piping for heating and cooling systems.
- calculate heat loss to design HVAC systems needed for specific situations.
- plan and draw a passive and active solar heating and cooling system.

USING THE COLOR IMAGE MASTERS AND IMAGE MASTERS

The following color image masters (CIM), image masters (IM), and handout masters (HO) are appropriate for use with this chapter. Teaching suggestions are included.

IM-32A: Solar Collectors

- Use to show a simplified version of how solar collectors work.

HO-32A: Hot-Water Heating Systems

- Use to show students the different hot-water pipe arrangements. Discuss the advantages and disadvantages of each.

HO-32B Insulation Use

- This illustration summarizes the role of insulation to retard heat and sound transfer.

EVALUATION

1. Assign the exercises at the end of the textbook chapter. When applicable, check students' answers against the answers given in the following section.
2. Test student mastery of factual material by using Chapter 32 Review found in the "Chapter Reviews" section of this guide.
3. In a check for competence, students should demonstrate understanding of:
   - the concepts for heating and cooling a house.
   - the concepts of heat transfer.
   - the different types of insulation.
   - the concepts of solar heating.
4. In a check for competence, students should be able to perform these skills:
   - plan and draw on a floor plan a warm-air heating system, a radiant heating system, a hot-water heating system, and a cooling system.

ANSWERS TO END-OF-CHAPTER EXERCISES

Many end-of-chapter exercises ask students to perform an activity, such as draw a floor plan or photograph local structures. As a result, no "answer" is required and none is given here.

1. Factors include the difference between indoor and outdoor temperatures, the type and thickness of building materials, the amount and type of insulation, and the amount of air leakage (infiltration) into or out of a structure.
2. No answer required.
3. Collecting, storing, distributing, controlling.
4. No answer required.
5. Solar furnace—collectors must move with the sun and are not currently practical for single building use.

   Photovoltaics—cells are not yet economically feasible for individual building use.

   Liquid-based systems—a cold climate requires a downdrain system; works well even on cold days; absorbers designed to retain maximum amount of heat.

   Air-based systems—simpler to build and operate than liquid-based systems; unable to provide hot water; low heat capacity requires large ducts.

6–9. No answers required.

10. Answers will vary.

Date	M	Tu	W	Th	F

Lesson Plan for Chapter 33 (Text pages 617–630)

Plumbing Drawings

ESSENTIAL ELEMENTS

For a complete listing of the essential elements, see the "Program Development" section in this guide. Essential elements for the course: 3.5, 5.3, 6.5, 6.6, 6.7, 6.9, 6.11, 6.15

CHAPTER 33 OUTLINE

Plumbing Conventions and Symbols
Plumbing Systems
    Water Supply System
    Waste Discharge System
Plumbing Drawings
    Plumbing Floor Plans
    Plan Details
    Plumbing Elevations

OBJECTIVES

Following are the objectives that appear at the beginning of Chapter 33 in the textbook. Students are expected to learn to:

- draw plumbing fixtures on a floor plan.
- draw the water supply lines and waste discharge system on a floor plan.
- draw the water supply lines and waste discharge system on an elevation.

USING THE COLOR IMAGE MASTERS AND IMAGE MASTERS

The following color image masters (CIM), image masters (IM), and handout masters (HO) are appropriate for use with this chapter. Teaching suggestions are included.

IM-33A: Plumbing Symbols

- Use to show symbols in both orthographic and isometric form.
- Make photocopies of this master and give to students for their notebooks.

HO-33A: Water Pressure System

- Use to show the basic components of a water pressure system. The plan and elevation drawings show how supply pipes appear on plumbing drawings.

HO-33B: Waste System

- The plan and elevation drawings show how a waste system appears on plumbing drawings. Ask students why all the pipes in this system slant downward and why they are larger than those in a water supply system.
- Point out that vents provide for air circulation, equalize pressure, and permit gases to escape. Fixture traps prevent gases from backing up into the building.

IM-33B Gravity Waste System

- A profile of a gravity waste system is shown here.

EVALUATION

1. Assign the exercises at the end of the textbook chapter. When applicable, check students' answers against the answers given in the following section.
2. Test student mastery of factual material by using Chapter 33 Review found in the "Chapter Reviews" section of this guide.

3. In a check for competence, students should demonstrate understanding of:
   - alternative types of plumbing fixtures.
   - how to order plumbing fixtures, including costs, colors, and materials.
   - concepts of pressure-water and gravity-waste systems.
   - pipe sizes for water pressure and waste systems.
   - different methods of heat distribution.
   - all the air-conditioning symbols used for a house plan.
4. In a check for competence, students should be able to perform these skills:
   - draw plumbing fixtures on a floor plan.
   - draw the hot and cold water-pressure lines on a floor plan.
   - draw the gravity waste system on a floor plan.
   - draw the symbols for insulation, mechanical equipment, and warm and cool air ducts on a floor plan.

ANSWERS TO END-OF-CHAPTER EXERCISES

Many end-of-chapter exercises ask students to perform an activity, such as draw a floor plan or photograph local structures. As a result, no "answer" is required and none is given here.

1. A water supply system consists of a network of pipes that carry fresh water under pressure to appliances, sinks, water closets (toilets), tubs, filters, showers, and water heaters.
2. Stacks and branches.

3–8. No answers required.

Date	M	Tu	W	Th	F

Lesson Plan for Chapter 34 (Text pages 631–643)

Drawing Coordination and Checking

ESSENTIAL ELEMENTS

For a complete listing of the essential elements, see the "Program Development" section in this guide. Essential elements for the course: 3.3, 3.5, 4.3, 5.3, 6.4, 6.5, 6.6, 6.7, 6.15

CHAPTER 34 OUTLINE

Checking Drawings
- Checking Methods
- Agreement Among Drawings
- Checking Sets of Drawings

Corrections and Changes
- Change Orders
- Changes on Drawings

OBJECTIVES

Following are the objectives that appear at the beginning of Chapter 34 in the textbook. Students are expected to learn:

- to organize and check a complete set of architectural drawings.
- how drawings in a set are related.
- to identify identical locations on all drawings in a set.
- to select the drawings needed to complete a set of architectural drawings.
- methods of drawing and recording changes on drawings according to change orders.

USING THE COLOR IMAGE MASTERS AND IMAGE MASTERS

The following color image masters (CIM), image masters (IM), and handout masters (HO) are appropriate for use with this chapter. Teaching suggestions are included.

IM-34A: Callouts

- Point out in the top drawing how a callout symbol identifies the section drawing sheet location and the sheet which includes the cutting plane line.
- Point out in the bottom drawing how titles are used to identify the source of the drawing and the scale.

HO-34A: Drawing Labels

- Use to show how cutting plane lines attached to section callout symbols identify the detail section number and drawing sheet. The bottom illustration shows how section details are labeled to identify the source of the detail by sheet and detail number.

HO-34B Basic Set of Plans

- The plot plan, floor plan and elevations of a basic plan are shown here to provide a comparison among drawing in a set. Have students locate similar positions on each drawing.

IM-34B Elevation Section

- This is the elevation section relating to the plan set included in HO-34B.

IM-34C Stair Section

- This is a stair section relating to the plan set in HO-34B.

IM-34D Fireplace Plan

- This is a fireplace plan which is part of the set of plans found in HO-34B.

IM-34E Stair Plan

- This stair plan is part of the set of plans in HO-34B.

EVALUATION

1. Assign the exercises at the end of the textbook chapter. When applicable, check students' answers against the answers given in the following section.
2. Test student mastery of factual material by using Chapter 34 Review found in the "Chapter Reviews" section of this guide.
3. In a check for competence, students should demonstrate understanding of:
   - the sequence for creating a set of drawings.
   - the geometrical symbols on the set of plans.
   - the system code used on drawings.
   - how much detail a particular set of plans requires.
   - all the drawings listed on a checklist.
4. In a check for competence, students should be able to perform these skills:
   - create a specific drawing and detail for each area in a set of plans.
   - locate identical points on all parts of the plan.
   - draw both a minimum and a typical set of plans.
   - select the number and type of drawings needed to complete a set of drawings from a checklist.

ANSWERS TO END-OF-CHAPTER EXERCISES

Many end-of-chapter exercises ask students to perform an activity, such as draw a floor plan or photograph local structures. As a result, no "answer" is required and none is given here.

1. Structural design, materials, projections of views, indexing among views (cross referencing), symbols, changes, legibility, title blocks, dimensions and notes, agreement among drawings, code compliance.
2. Answers will vary.
3. No answer required.
4. Answers will vary.
5. The client, architect, contractor, or subcontractor.
6. A description of the change, the adjustment in costs, and a print of the altered design or correction.
7. A number inside a triangle.

Date	M	Tu	W	Th	F

Lesson Plan for Chapter 35 (Text pages 644–657)

Schedules and Specifications

ESSENTIAL ELEMENTS

For a complete listing of the essential elements, see the "Program Development" section in this guide. Essential elements for the course: 3.2, 3.4, 3.5, 5.3, 6.6, 6.7, 6.8, 6.9, 6.13, 6.15

CHAPTER 35 OUTLINE

Schedules
- Window Schedules
- Door Schedules
- Hardware and Fittings
- Plumbing Fixtures
- Electrical Fixture Schedules
- Appliances
- Surface Coverings Schedules
- Paint and Finishing Schedules
- Furniture and Accessory Schedules
- Construction Component Schedules

Material Lists

Specifications
- Purpose and Use
- Guidelines for Writing Specifications
- Organization
- Reference Sources

OBJECTIVES

Following are the objectives that appear at the beginning of Chapter 35 in the textbook. Students are expected to learn to:

- create schedules for a set of architectural drawings.
- develop a material list for on-site construction.
- create a set of specifications in a standard format.

USING THE COLOR IMAGE MASTERS AND IMAGE MASTERS

The following color image masters (CIM), image masters (IM), and handout masters (HO) are appropriate for use with this chapter. Teaching suggestions are included.

IM-35A: Simplified Door and Window Callouts

- Point out that simplified callouts make reference to door or window schedules for further information.

HO-35A Schedule Information

- This handout lists the major items to be included in a window, door and closet schedule.

EVALUATION

1. Assign the exercises at the end of the textbook chapter. When applicable, check students' answers against the answers given in the following section.
2. Test student mastery of factual material by using Chapter 35 Review found in the "Chapter Reviews" section of this guide.
3. In a check for competence, students should demonstrate understanding of:
   - the efficiency of door and window styles.
   - alternate styles of windows and doors.
   - different types of paints and finish materials.
   - alternative types of appliances and fixtures.
   - the approximate costs of all items used on a plan.
   - all areas of a list of specifications.
   - the form used for a contractor's materials list.

4. In a check for competence, students should be able to perform these skills:
   - make up door and window schedules for a floor plan.
   - make up paint and materials schedules for a floor plan.
   - make up appliance and fixture schedules for a floor plan.
   - create a list of specifications for a house.

ANSWERS TO END-OF-CHAPTER EXERCISES

Many end-of-chapter exercises ask students to perform an activity, such as draw a floor plan or photograph local structures. As a result, no "answer" is required and none is given here.

1. Product or component information; referenced with a key symbol.
2. No answer required.
3. To ensure that a building will be constructed as designed; to prevent misunderstandings among the client, the architect, and the builder; by lending institutions to evaluate the quality of construction; by contractors to make estimates and bids.
4. Schedules are charts that contain product or component information; specifications are written descriptions of what materials are to be used and how; materials lists are lists of materials to be used at the building site.
5. Division, section, subsection, and detail; detail added using subdivisions of five-digit code.
6. Division—wood and plastics; section—heavy timber.
7. Manufacturer's literature, trade association publications, compilations of information such as *Sweets Catalog Files*.

8–10. Answers will vary.

Date	M	Tu	W	Th	F

Lesson Plan for Chapter 36 (Text pages 658–664)

Building Costs and Financial Planning

ESSENTIAL ELEMENTS

For a complete listing of the essential elements, see the "Program Development" section in this guide. Essential elements for the course: 3.1, 3.2, 5.2, 5.3, 6.8, 6.13, 6.15

CHAPTER 36 OUTLINE

Building Costs
- Cost Estimating Methods
- Minimizing Costs

Financial Planning
- Mortgage
- Interest
- Taxes
- Insurance
- Soft Costs and Closing Costs
- Budgets and Financing Qualifications

OBJECTIVES

Following are the objectives that appear at the beginning of Chapter 36 in the textbook. Students are expected to learn to:

- estimate building costs by the square-foot method.
- estimate building costs by cubic volume.
- make up a home budget.
- calculate monthly payments.

USING THE COLOR IMAGE MASTERS AND IMAGE MASTERS

The following color images masters (CIM), image masters (IM), and handout masters (HO) are appropriate for use with this chapter. Teaching suggestions are included.

HO-36A: House Design Reality Check

- Use these worksheets to help students understand the costs involved in building a home today. The point is not to destroy their dreams but to help them work toward those dreams in realistic ways.

EVALUATION

1. Assign the exercises at the end of the textbook chapter. When applicable, check students' answers against the answers given in the following section.
2. Test student mastery of factual material by using Chapter 36 Review found in the "Chapter Reviews" section of this binder.
3. In a check for competence, students should demonstrate understanding of:
   - the concepts of mortgage and interest payments.
   - how to shop for less expensive loans.
   - the breakdown of costs for land and home construction.

- how to cut costs.
- closing costs.

4. In a check for competence, students should be able to perform these skills:
   - make up a home budget.
   - compute monthly payments.
   - estimate building costs by the square foot and by cubic volume.
   - figure landscaping costs.

ANSWERS TO END-OF-CHAPTER EXERCISES

Many end-of-chapter exercises ask students to perform an activity, such as draw a floor plan or photograph local structures. As a result, no "answer" is required and none is given here.

1. $84,000
2. $86,400
3. Answers will vary.
4. $0
5. $954
6. Answers will vary.
7. Monthly mortgage payments.
8. $298
9. Answers will vary.

Date	M	Tu	W	Th	F

Lesson Plan for Chapter 37 (Text pages 665–668)

Codes and Legal Documents

ESSENTIAL ELEMENTS

For a complete listing of the essential elements, see the "Program Development" section in this guide. Essential elements for the course: 3.1, 3.2, 3.3, 5.3, 6.8, 6.10, 6.15

CHAPTER 37 OUTLINE

Building Codes
- Types of Codes
- Code Compliance

Legal Documents
- Contracts
- Bonds
- Deeds
- Liens
- Construction Bids

OBJECTIVES

Following are the objectives that appear at the beginning of Chapter 37 in the textbook. Students are expected to learn to:

- consider building codes in architectural design.
- determine legal documents needed for building construction.

USING THE COLOR IMAGE MASTERS AND IMAGE MASTERS

The following color image masters (CIM), image masters (IM), and handout masters (HO) are appropriate for use with this chapter. Teaching suggestions are included.

IM-37A: Load Categories

- Point out that loads vary greatly, depending upon the type of structure. This illustration shows typical loads for a two-level, wood-frame building. Loads are given in pounds per square foot.
- Remind students that dead loads are permanent parts of a building. Live loads are variable, such as people, furniture, cars, and forces of nature.

HO-37A: Code Requirements

- Give to students to use as reference when writing requirements for structural members such as footings, lintel, and ceiling joist spans for light construction.

HO-37B: Joist and Rafter Tables

- Give to students to use as reference when calculating spans for light construction.

EVALUATION

1. Assign the exercises at the end of the textbook chapter. When applicable, check students' answers against the answers given in the following section.
2. Test student mastery of factual material by using Chapter 37 Review found in the "Chapter Reviews" section of this guide.
3. In a check for competence, students should demonstrate understanding of:
   - engineering tables.
   - how the local building department operates.
   - specialized areas having building regulations.
   - the concept of building loads.

4. In a check for competence, students should be able to perform these skills:
   - read local and state building regulations.
   - calculate dead and live loads on a building.

ANSWERS TO END-OF-CHAPTER EXERCISES

Many end-of-chapter exercises ask students to perform an activity, such as draw a floor plan or photograph local structures. As a result, no "answer" is required and none is given here.

1. Collections of laws that ensure that minimum building standards are met; they are enacted to safeguard life, health, property, and the public welfare.
2. Performance-oriented codes do not limit or specify the use of most construction methods or materials; they establish only safety and performance requirements. Specification codes include specific requirements for the use and location of materials and methods of construction.
3. Answers will vary.
4. Answers will vary.
5. Codes apply to all projects within a given area; legal agreements or contracts are created for each individual building project.
6. A bid form gives information to bidders regarding verification of receipt of all drawings and documents, specific length of time the bid will be open, price quotation for the bid, and a list of substitute materials or components if any item varies from specifications. When bidders sign the form, they agree to abide by all conditions of the bid.

Using the Student Workbook

INTRODUCTION

The *Drafting and Design for Architecture Student Workbook* supplements the student text by providing drawing exercises keyed to chapter content. This section provides tips for using the exercises and tutorials and an answer key for exercises that require specific responses.

USING THE DRAWING EXERCISES

The Student Workbook includes 91 drawing exercises keyed to chapters in the student text. The exercises will help students strengthen their knowledge and understanding of architectural fundamentals. Completing the exercises before making drawings of their own design will help clarify concepts and procedures, and they will avoid time-consuming errors.

The exercises can be assigned as pretests, post-tests, homework assignments, extra-credit assignments, classroom activity and discussion assignments, and introductions to topics you wish to cover in class. You might also allow individual students to choose to complete those exercises involving subjects or skills in which they are especially interested.

Although the organization of the exercises follows that of the text and are designed to reinforce text material, the exercises may be used in any sequence. However, they will naturally be more effective if students have already covered any prerequisite subjects or skills.

The exercises may be done using manual drafting and sketching techniques or with a computer-aided drafting (CAD) system. They are flexible enough to be appropriate for most software programs. However, we recommend that students develop a solid working knowledge of manual sketching and drafting before moving on to computer work. The drawing exercises provide a good means for practicing that knowledge.

Following the exercises are five sheets of drawing templates. Some of the templates are required for specific exercises. Others are for students to use for their own drawings, as desired.

A few of the exercises involve the student in choosing right or wrong answers or solutions. An answer key for these exercises can be found later in this section of the Teacher's Resource Binder. The drawing activities have no right or wrong answer, and student approaches will vary. Encourage your students to be creative in their thinking and designs. Do not judge their design solutions until you have given them an opportunity to defend them either in a discussion with you or in a written report.

Because construction methods vary considerably in different locations, there may be many acceptable solutions to any construction problem. You should stress this point to students and have them use the most common methods found in your own community.

ANSWER KEY FOR STUDENT WORKBOOK

The Student Workbook consists primarily of drawing exercises; therefore, few right or wrong answers are required. When they are, or when a particular drawing needs a specific response, the answers are given here.

Exercise 2-1

The house has asymmetrical balance.

Exercise 3-1

Mechanical Engineer's Scale (from top down)

1. 3 11/16″
2. 3 3/16″
3. 2 5/8″
4. 2 3/16″
5. 1 1/4″
6. 1/2″

Architect's Scale (from top down)

1. 13′-11″
2. 10′-11″
3. 6′-7″
4. 4′-3″
5. 3′
6. 1′

Exercise 3-2

Make a template or overlay to check students' answers, or use an architect's scale.

Exercise 5-1

DATA TABLE					
Line	X	Y	go to >	X	Y
1	1	1	>	1	12
2	1	12	>	11	12
3	11	12	>	22	16
4	22	16	>	24	11
5	24	11	>	18	1
6	18	1	>	1	1

Exercise 5-2

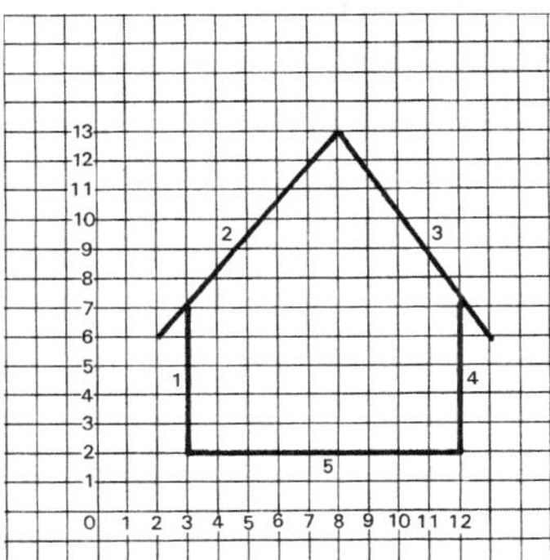

Exercise 5-3

Line	X	Y	go to —>	X	Y
1	1	9	—>	19	9
2	19	9	—>	19	1
3	19	1	—>	15	1
4	15	1	—>	15	4
5	15	4	—>	12	7
6	12	7	—>	1	7

Line	X	Y	go to —>	X	Y
1	1	18	—>	19	18
2	19	18	—>	19	5
3	19	5	—>	21	5
4	21	5	—>	21	2
5	21	2	—>	14	2
6	14	2	—>	14	5
7	14	5	—>	16	5
8	26	5	—>	16	14
new start					
9	16	13	—>	19	15
new start					
10	1	14	—>	19	14

Exercise 5-4

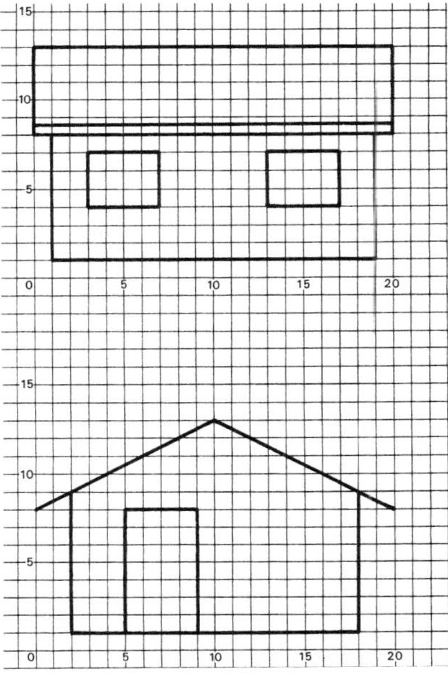

Exercise 9-1

Service Circulation	Approximate distance	Excellent	Fair	Poor
Service door to kitchen	5'		√	
Service door to living area	40'		√	
Service door to bathroom	6'	√		
Service door to bedrooms	55'			√
Kitchen to bathrooms	30'		√	
Paths around work triangle	20'	√		

Guest's Circulation	Approximate distance	Excellent	Fair	Poor
Front door to closet	2'	√		
Front door to living room	10'	√		
Front door to bathroom	20'		√	
Front door to outdoor living	25'		√	
Living room to bathroom	35'		√	

Resident's Circulation	Approximate distance	Excellent	Fair	Poor
Front door to kitchen	30'		√	
Kitchen to bedrooms	45'			√
Kitchen to children's play area	10'	√		
Kitchen to children's sleeping area	45'			√
Bedrooms to bathrooms	5'	√		
Outdoor living area to bathroom	20'		√	
Living room to outdoor living area	2'	√		

Exercise 9-2

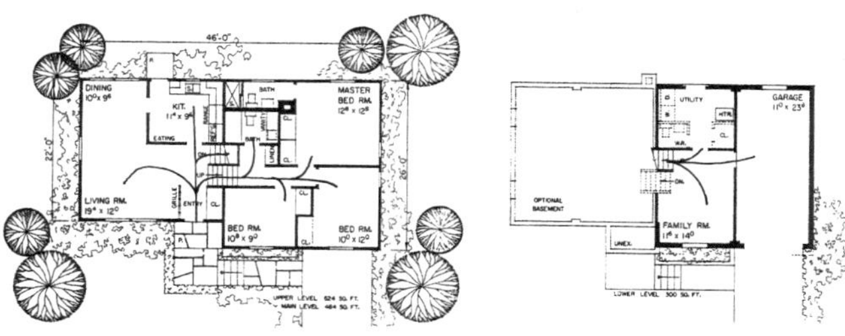

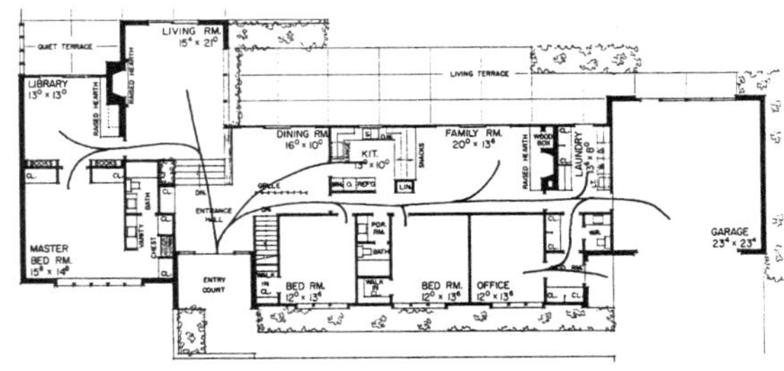

Exercise 18-3

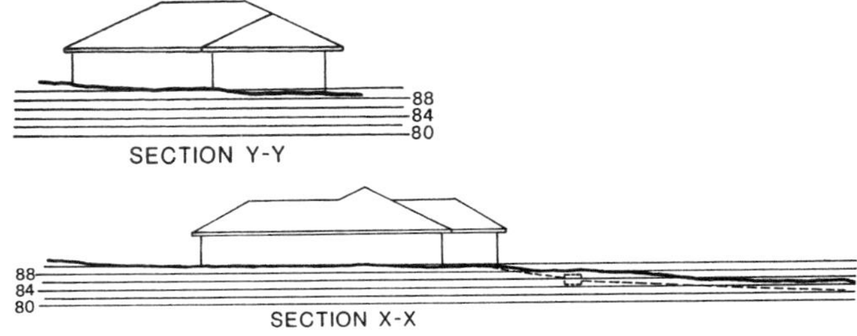

Exercise 28-2

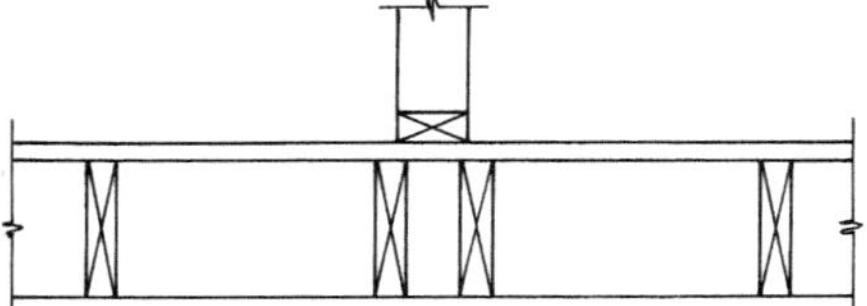

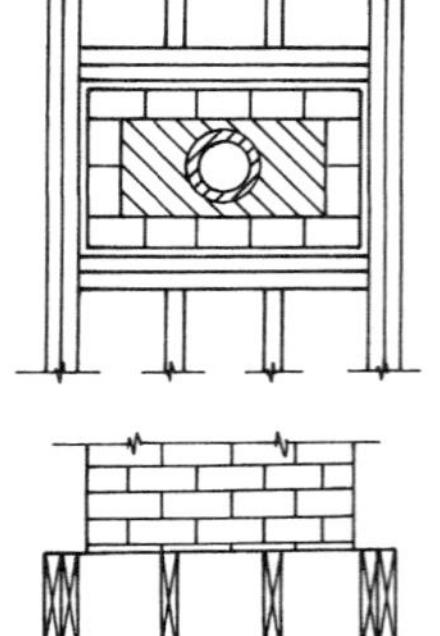

Exercise 29-1

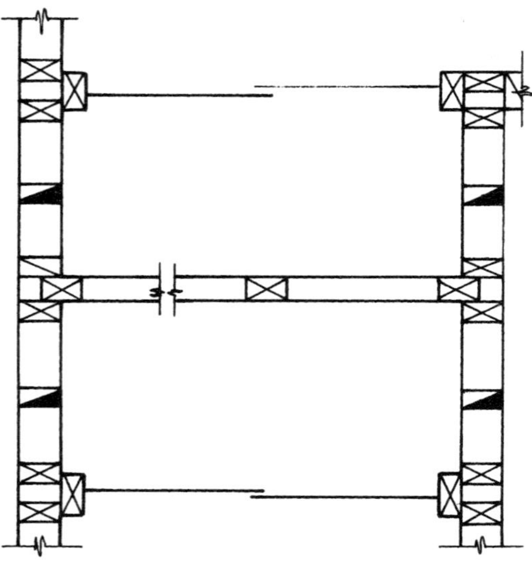

Exercise 29-2

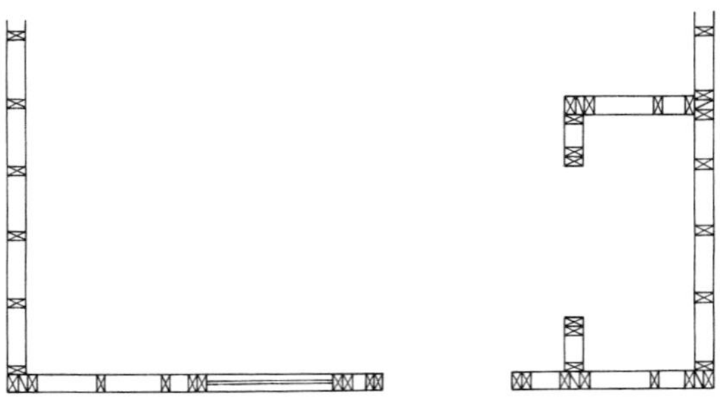

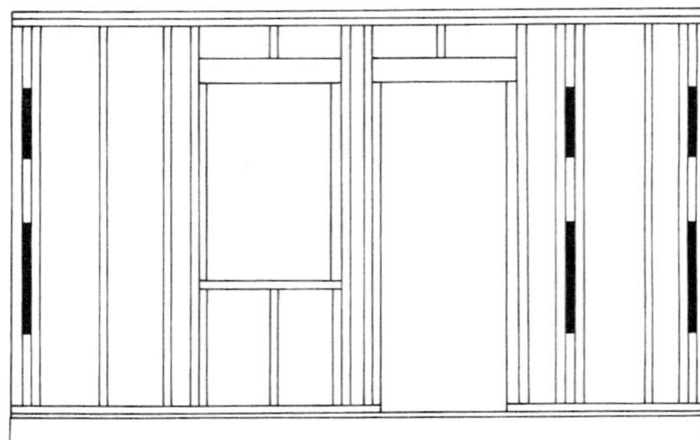

Exercise 30-3

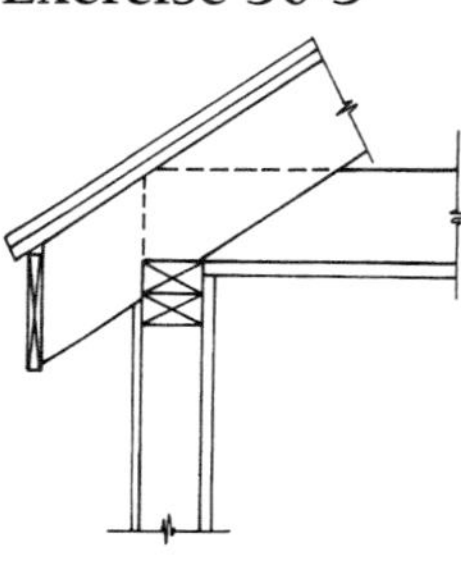

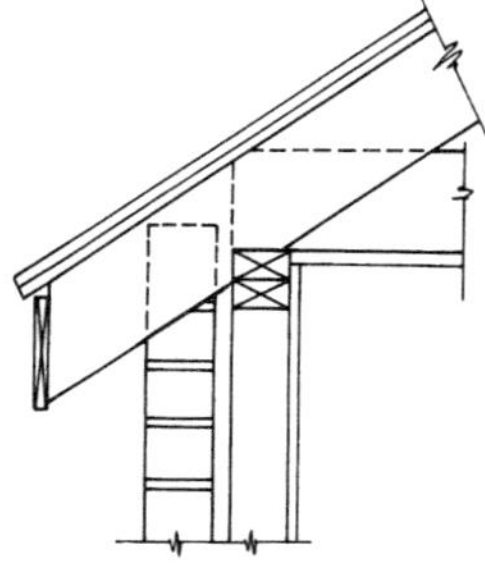

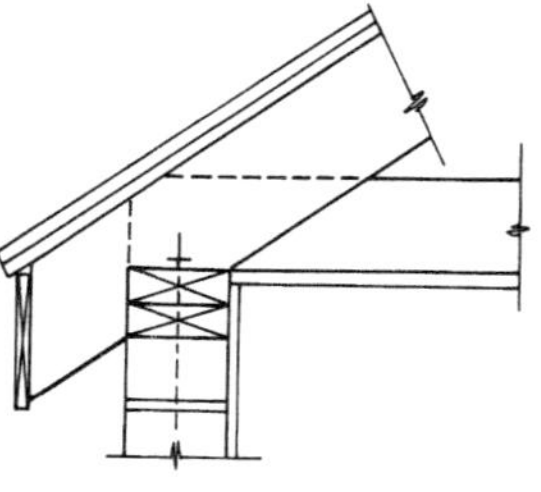

Exercise 30-4

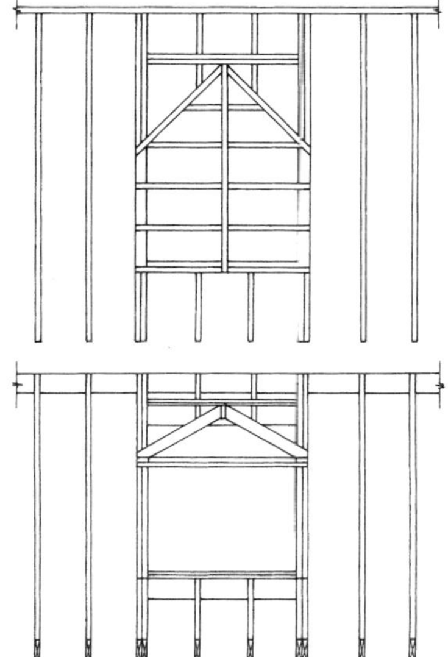

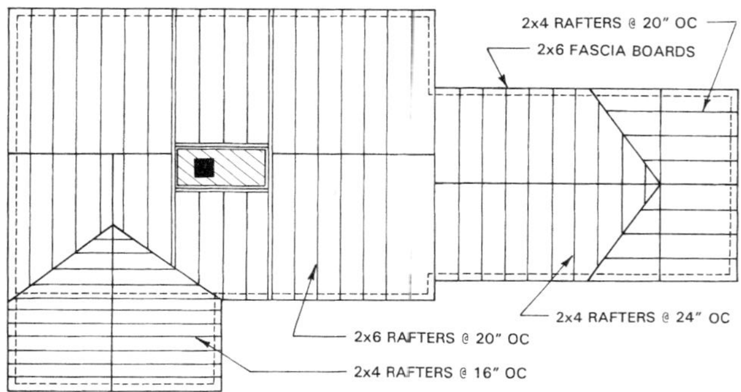

Exercise 34-1

- no door in dining room wall
- no dashed lines connecting switches to fixtures
- master bedroom name is incomplete
- left wall dimension 13′-7″ should be 14′-9″
- top wall dimension 33′-8″ should be 43′-8″
- no sink in kitchen
- no flue in fireplace
- doorbell button in railing, not wall
- no GFCI in kitchen

Exercise 36-1

Building costs $75/sq. ft.
Square Footage—1200
Total Cost—$90,000

Building costs $90/sq. ft.
Square Footage—1400
Total Cost—$126,000

Building costs $95/sq. ft.
Square Footage—2000
Total Cost—$190,000

Building costs $100/sq. ft.
Square Footage—2856
Total Cost—$285,600

Exercise 36-2

1. $39,000
2. $58,500
3. $122,700
4. 14,850 cu. ft.
   21,262.5 cu. ft.
5. $14,895
6. $1319
7. 68%
8. $100,000

Exercise 37-1

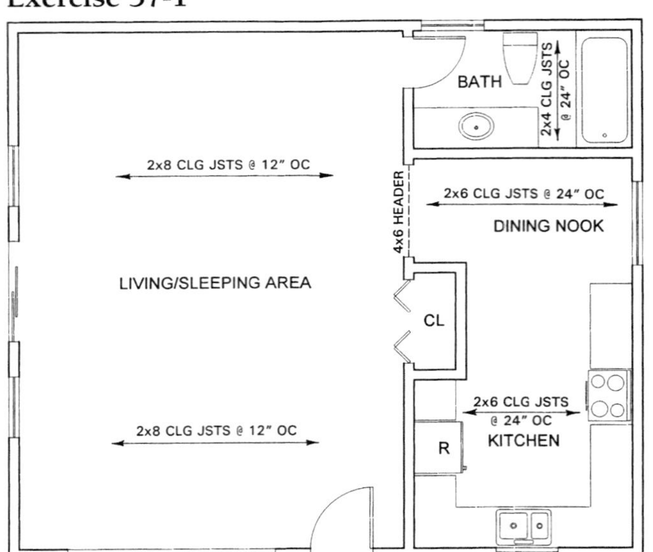

Exercise 37-2

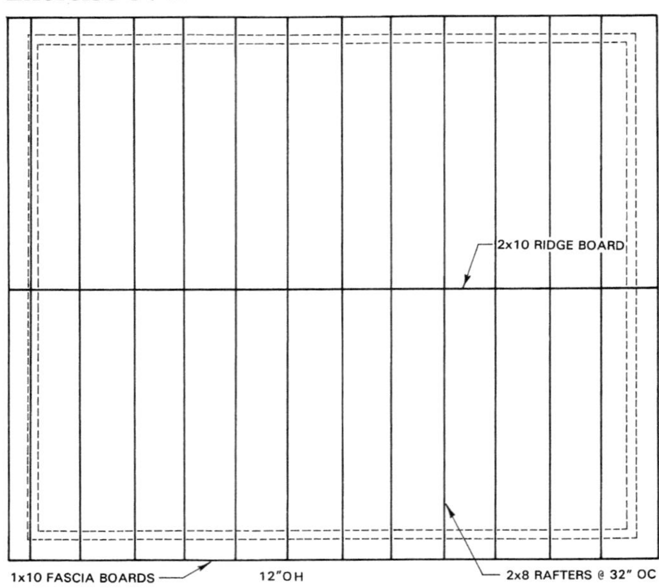

Interpreting Architectural Drawings Tests

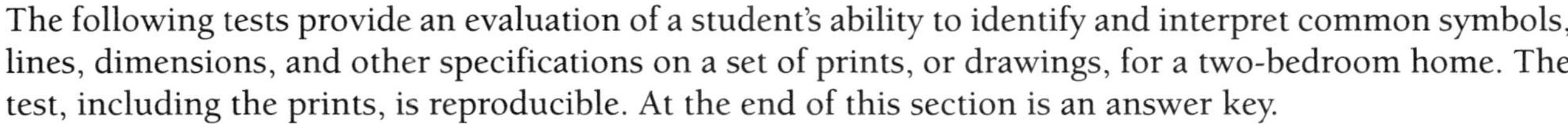

The following tests provide an evaluation of a student's ability to identify and interpret common symbols, lines, dimensions, and other specifications on a set of prints, or drawings, for a two-bedroom home. The test, including the prints, is reproducible. At the end of this section is an answer key.

The prints include:

Drawing 1: Floor Plan
Drawing 2: Plot Plan
Drawing 3: East and West Exterior Elevations and Roof Plan
Drawing 4: North and South Exterior Elevations
Drawing 5: Electrical and Perimeter Radial Heating Plan
Drawing 6: Front Landing Plan and Window and Door Schedules
Drawing 7: Framing and Interior Elevations
Drawing 8: Slab Foundation Plan
Drawing 9: T-Foundation Plan
Drawing 10: Construction Details (foundation, fireplace, and eaves)
Drawing 11: Garage and Roof Framing Plans

An additional way to use the drawings is to add letters to specific locations on the floor plan and have students place letters on corresponding locations on the other drawings.

Name________________________ Class ____________ Date ____________

INTERPRETING ARCHITECTURAL DRAWINGS TESTS

The following questions are keyed to a set of 11 prints for a two-bedroom house. Be sure you are looking at the right print for each list of questions. Write your answer to each question in the blank at the left.

Drawing 1: Floor Plan

______________________ 1. What is the size of the ceiling joists?

______________________ 2. What is the spacing of the ceiling joists?

______________________ 3. What does the abbreviation "OC" stand for?

______________________ 4. What is the width of the front door?

______________________ 5. What does the abbreviation "LAM BM" stand for?

______________________ 6. What type of door is used for the closet in the front entry?

______________________ 7. What does the abbreviation "FAU" stand for?

______________________ 8. Draw the symbol for an elevation callout.

______________________ 9. What type of line is used to indicate the wall cabinets?

______________________ 10. What does the abbreviation "WH" stand for?

______________________ 11. Of what type is door number 3?

______________________ 12. What does the abbreviation "D" stand for?

______________________ 13. What does the abbreviation "W" stand for?

______________________ 14. What does the abbreviation "LT" stand for?

______________________ 15. What does the abbreviation "R" stand for?

______________________ 16. What does the square black symbol in the fireplace represent?

______________________ 17. What do the 45° lines in the fireplace represent?

______________________ 18. What are the overall dimensions of the house?

______________________ 19. What is used to hang the ceiling joists?

______________________ 20. What callout is used for the living room windows?

______________________ 21. What is the total square footage of the house?

______________________ 22. What scale is indicated on the floor plan?

______________________ 23. What geometric symbol is used for a door callout?

______________________ 24. What geometric symbol is used for a window callout?

______________________ 25. What structural member supports the ceiling joists in the living room and dining room?

Name________________________________ Class ____________ Date ____________

Drawing 2: Plot Plan

________________ 26. What scale is indicated on the plot plan?

________________ 27. What material is used for the front landing?

________________ 28. In what direction is the front of the house facing?

________________ 29. What is the overall size of the double garage?

________________ 30. What is the front setback of the house?

________________ 31. What is the rear setback of the house?

________________ 32. What is the rear setback of the garage?

________________ 33. What are the overall dimensions of the site?

________________ 34. What is the width of the front walk?

________________ 35. What do the dotted lines represent?

Drawing 3: East and West Exterior Elevations and Roof Plan

________________ 36. What does the abbreviation "CONT OH" stand for?

________________ 37. What scale is indicated on the roof plan?

________________ 38. What does the abbreviation "GA GI" stand for?

________________ 39. What is used to cover the ends of the rafters?

________________ 40. What is the triangle having units of 12 and 6 called?

________________ 41. What is the pitch of the roof?

________________ 42. What do the arrows in the middle of the roof plan describe?

________________ 43. What is the roof type?

________________ 44. What scale is indicated on the elevation drawings?

________________ 45. What do the dotted lines on the roof plan represent?

________________ 46. What does the small rectangle at the bottom of the roof represent?

________________ 47. What exterior material is used on the fireplace?

________________ 48. What type of window is installed in the west elevation?

________________ 49. What type of window is installed in the east elevation?

________________ 50. What do the dotted lines above the windows in the elevations indicate?

Name________________________ Class ____________ Date ____________

Drawing 4: North and South Exterior Elevations

________________ 51. What is the standard window height?

________________ 52. What is the standard door height?

________________ 53. What is the ceiling height?

________________ 54. What is the chimney clearance from the ridge board?

________________ 55. What do the dotted lines on windows and doors indicate?

________________ 56. What do the arrowheads on windows indicate?

________________ 57. Which elevation faces the backyard?

________________ 58. What scale is indicated on the elevation drawings?

________________ 59. In the south elevation, on which side of the door are the hinges placed?

________________ 60. What type of foundation is used for the elevations?

Drawing 5: Electrical and Perimeter Radial Heating Plan

________________ 61. Draw the symbol for a heat outlet.

________________ 62. Draw the symbol for an electrical floor outlet.

________________ 63. Draw the symbol for a switch.

________________ 64. Draw the symbol for a convenience duplex outlet.

________________ 65. Draw the symbol for a split convenience duplex outlet.

________________ 66. Draw the symbol for a telephone outlet.

________________ 67. Draw the symbol for an electric wall heater.

________________ 68. What does the abbreviation "GFCI" stand for?

________________ 69. What type of heating system is used?

________________ 70. Draw the symbol for a three-way switch.

________________ 71. Draw the symbol for an electrical power panel.

________________ 72. Draw the symbol for a dimmer switch.

________________ 73. Draw the symbol for a 240V outlet.

________________ 74. Draw the symbol for a water heater.

________________ 75. How are the lights in the bedroom closets controlled?

________________ 76. Why would a switch be wired into a convenience outlet?

________________ 77. Draw the electrical symbol required for the cooking range.

Name____________________ Class __________ Date __________

_______________ 78. What is the diameter of the heating ducts?

_______________ 79. What is the material used for the heating ducts?

_______________ 80. When is a switch with a pilot light used?

Drawing 6: Front Landing Plan and Window and Door Schedules

_______________ 81. What is used to support the step?

_______________ 82. What connects the step to the landing?

_______________ 83. What is the size of the decking material?

_______________ 84. What is the size of the material making up the tread?

_______________ 85. What are the overall dimensions of the landing?

_______________ 86. What supports the landing against the house?

_______________ 87. What supports the outside edge of the landing?

_______________ 88. What is used to support the decking?

_______________ 89. What is the catalog number for the louvered window?

_______________ 90. What is the catalog number for door 2?

_______________ 91. What type of glass is used in window C?

_______________ 92. What are the dimensions for window D?

_______________ 93. Where will the solid core doors be located?

_______________ 94. What is the thickness of door 1?

_______________ 95. What is the finish for door 1?

Drawing 7: Framing and Interior Elevations

_______________ 96. What are the sizes of the studs?

_______________ 97. What is the spacing of the studs?

_______________ 98. What is the size of the window lintels?

_______________ 99. What is the size of the door lintel?

_______________ 100. What type of diagonal brace is used?

_______________ 101. How many studs are used to frame an exterior wall corner?

_______________ 102. How many studs are used to frame intersecting interior walls?

_______________ 103. In interior elevation 2, what is the built-in item on the left side of the fireplace?

Name________________________ Class ____________ Date ____________

__________________ 104. What type of window opening is used in interior elevation 2?

__________________ 105. What type of window opening is used in interior elevation 1?

__________________ 106. What is the opening in the lower left wall of interior elevation 3?

__________________ 107. What are the short studs under a window called?

__________________ 108. What are the short studs over the 4″ × 6″ lintel called?

__________________ 109. What is used to create a space between the studs at intersecting walls?

__________________ 110. What is the item in the right wall of interior elevation 3?

Drawing 8: Slab Foundation Plan

__________________ 111. Must the slab's overall dimensions be the same as those shown on the stud layout plan?

__________________ 112. What does section D-D show?

__________________ 113. What does section E-E show?

__________________ 114. What does section F-F show?

__________________ 115. What is the width of the interior footings?

__________________ 116. What is the thickness of the slab?

__________________ 117. What is the diameter of the slab's reinforcing bars?

__________________ 118. What is the spacing of the slab's reinforcing bars?

__________________ 119. What is the purpose of the slab's reinforcing bars?

__________________ 120. What scale is indicated on the slab foundation plan?

Drawing 9: T-Foundation Plan

__________________ 121. What is the size of the floor joists?

__________________ 122. What is the spacing of the floor joists?

__________________ 123. What is the spacing of the nonbearing piers?

__________________ 124. What is the spacing of the bearing piers?

__________________ 125. What does section A-A show?

__________________ 126. What does section B-B show?

__________________ 127. What does section C-C show?

__________________ 128. What is the purpose of double floor joists?

__________________ 129. What is the purpose of solid blocking?

Name__________________ Class __________ Date __________

__________ 130. What is the size of the girders?

__________ 131. What is the structural purpose of the girders?

__________ 132. What is the maximum span for this floor joist system?

__________ 133. What is the maximum spacing of the girders?

__________ 134. What structural members placed parallel to the floor joists are under the interior bearing walls?

__________ 135. What structural members placed perpendicular to the floor joists are under the bearing walls?

Drawing 10: Construction Details

__________ 136. What mechanical device stops cold air from entering the house through the fireplace flue?

__________ 137. What fireproof material lines the inside of the firebox?

__________ 138. Of what material is the hearth made?

__________ 139. To what structural members are the bottoms of the studs attached?

__________ 140. What is the clearance of the mud sill from the soil?

__________ 141. What is the depth of the slab's exterior footing?

__________ 142. What covers the roof?

__________ 143. What is the size of the fascia?

__________ 144. In the fireplace, what supports the top of the firebox opening?

__________ 145. What is the size of the steel reinforcing bars used in the fireplace?

__________ 146. From what material is the finished floor made?

__________ 147. What structural members are used to frame the opening in the floor for the fireplace?

__________ 148. What is the diameter of the anchor bolts used in the T-foundation?

__________ 149. What is the first structural wood member used on the T-foundation?

__________ 150. What is the width of the T-foundation footing?

__________ 151. What is the width of the T-foundation's wall?

__________ 152. What size reinforcing bars are used in the slab foundation?

__________ 153. What size reinforcing bars are used in the slab footing?

__________ 154. What is the inside liner of the chimney called?

__________ 155. What is the height of the opening in the fireplace?

Name________________________ Class ____________ Date ____________

________________ 156. What are the dimensions of the chimney opening?

________________ 157. What material is used to stop moisture from passing through an exterior wall?

________________ 158. What structural materials make up the top plate?

________________ 159. What is used to cover the ends of the rafters in the eaves?

________________ 160. What is the spacing of the roof rafters?

________________ 161. What is the spacing of the studs?

________________ 162. What is the spacing of the floor joists?

________________ 163. What is the spacing of the anchor bolts?

________________ 164. What is the thickness of the firebrick?

________________ 165. What structural members support the floor joists within the T-foundation?

________________ 166. What material is used for a vapor barrier in the roof construction?

________________ 167. How thick is the subfloor?

________________ 168. What material is used to seal the level of the mud sill?

________________ 169. How high above the hearth is the mantel?

________________ 170. How are the ends of the girders supported in the fireplace footing?

________________ 171. What scale is indicated on the fireplace detail drawing?

________________ 172. What scale is indicated on the eave detail drawing?

________________ 173. What is the depth of the fireplace footing?

________________ 174. What is the depth of the fill under the slab?

________________ 175. What structural material is used to close in the roof and walls?

Drawing 11: Garage and Roof Framing Plan

________________ 176. What is the size of the roof rafters?

________________ 177. What is the spacing of the roof rafters?

________________ 178. What is the size of the ridge board?

________________ 179. What materials are used in the roof to frame an opening for the chimney? (two answers)

________________ 180. How wide is the roof overhang?

________________ 181. What do the dotted lines in the roof plan represent?

________________ 182. On the garage plan, how long is the apron?

Name____________________________ Class ____________ Date ____________

____________________ 183. How thick is the driveway?

____________________ 184. What does the abbreviation "WWM" stand for?

____________________ 185. What is the width of the garage door?

____________________ 186. What is the width of the footing in the garage foundation?

____________________ 187. How thick is the garage slab?

____________________ 188. What is the depth of the fill used under the garage slab?

____________________ 189. The garage slab measures how many square feet?

____________________ 190. What size steel dowels connect the apron to the garage slab?

____________________ 191. What is the spacing of the steel dowels in the apron?

____________________ 192. What scale is indicated on the roof plan?

____________________ 193. What scale is indicated on the slab foundation detail drawing?

____________________ 194. What scale is indicated on the garage plan?

____________________ 195. What is the size of the garage window?

____________________ 196. What is the size of the swinging door in the side of the garage?

____________________ 197. What is the size of the steel dowels in the garage slab footing?

____________________ 198. How many continuous steel dowels are in the garage slab footing?

____________________ 199. Which section detail shows the apron?

____________________ 200. What is the apron's slope along its 2′-9″ length?

DRAWING 1: FLOOR PLAN

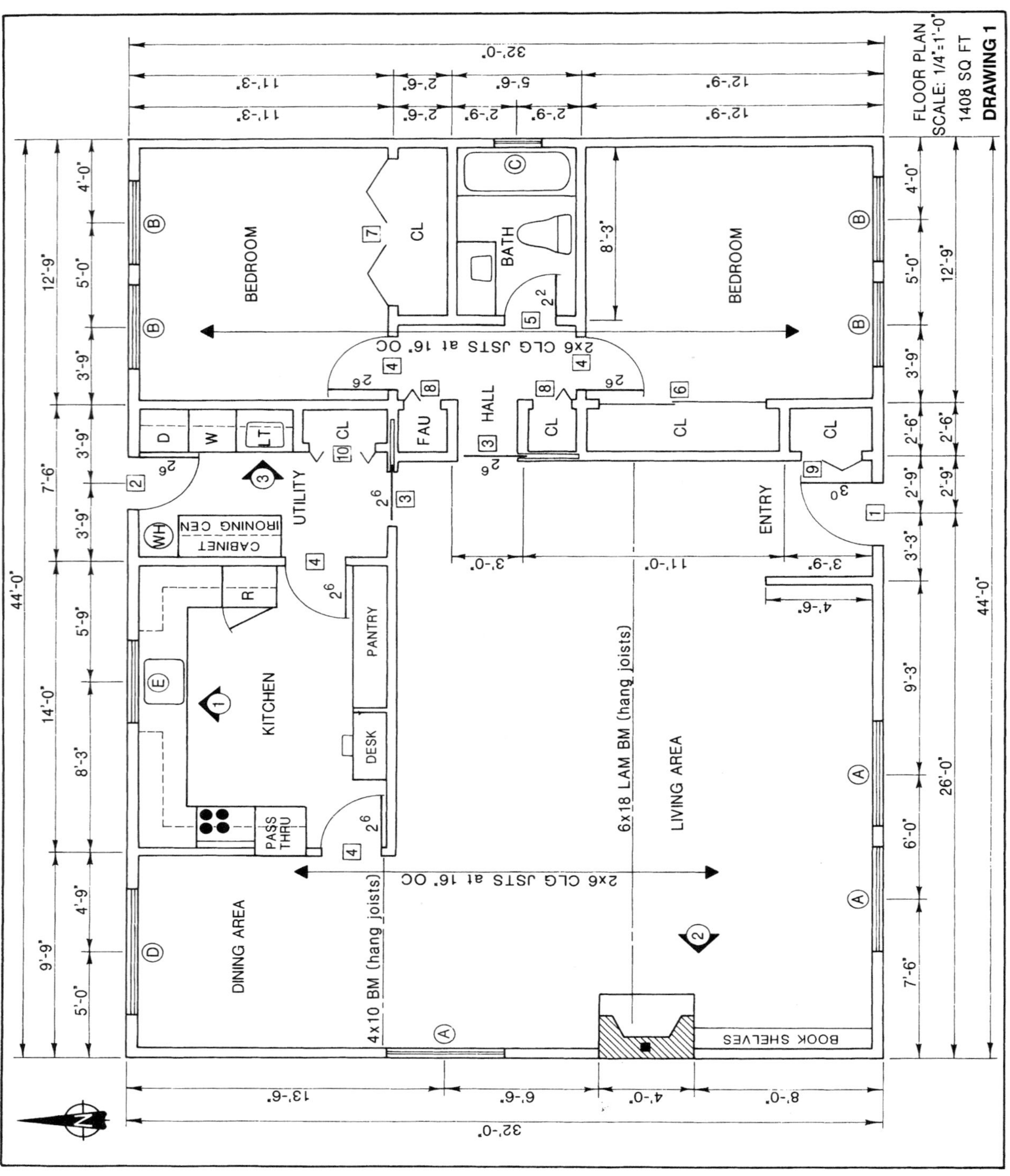

DRAWING 2: PLOT PLAN

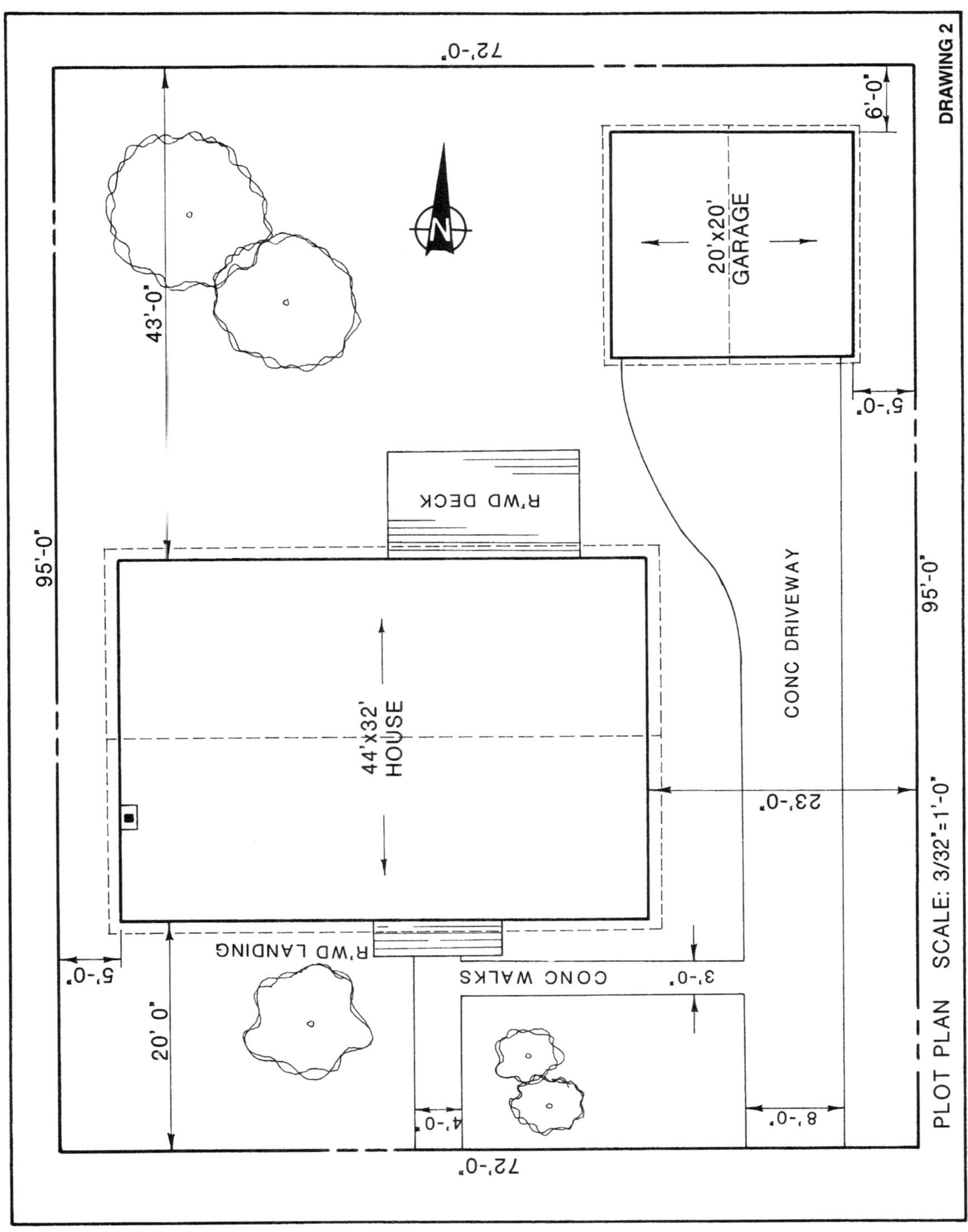

DRAWING 3: EAST AND WEST EXTERIOR ELEVATIONS AND ROOF PLAN

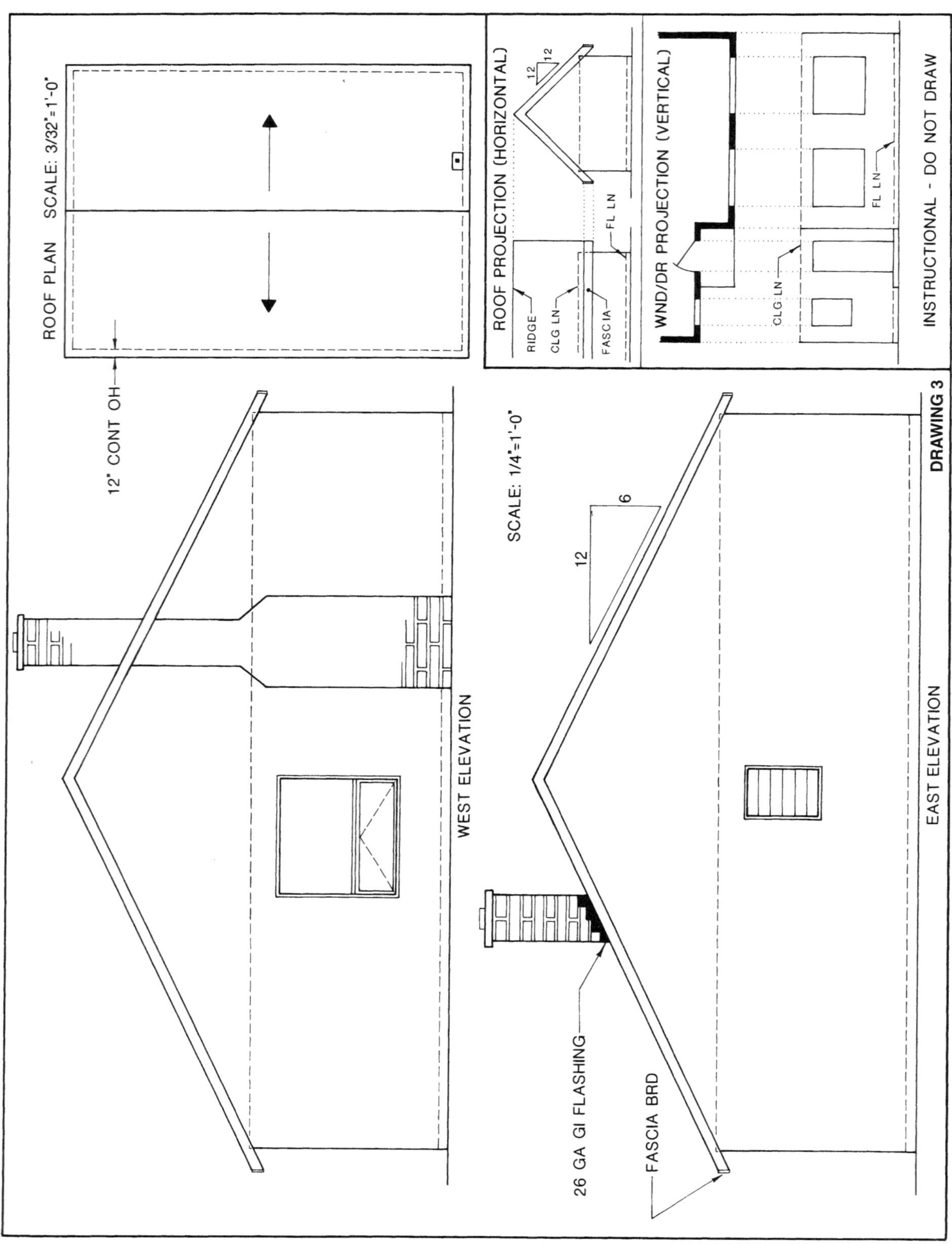

DRAWING 4: NORTH AND SOUTH EXTERIOR ELEVATIONS

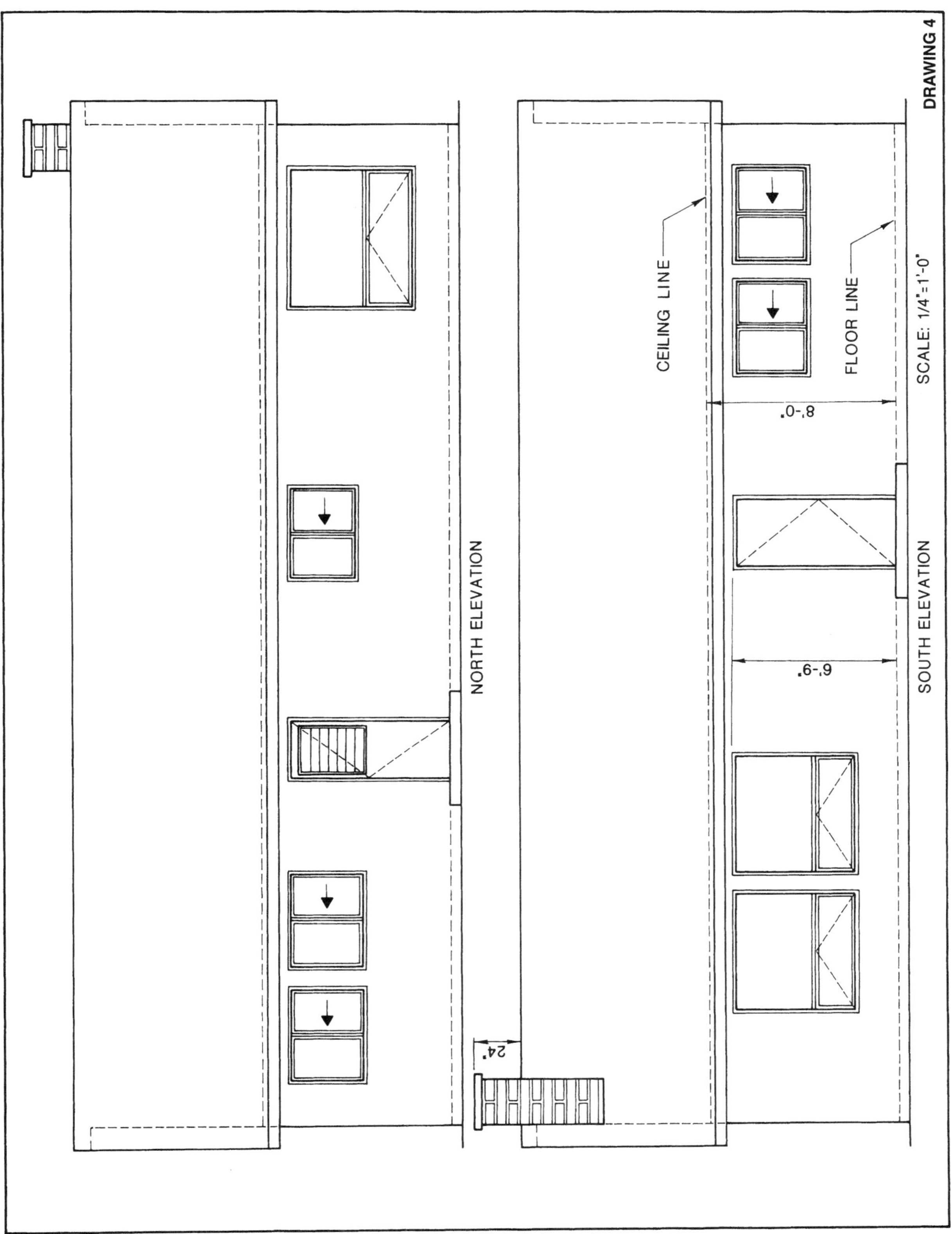

DRAWING 5: ELECTRICAL AND PERIMETER RADIAL HEATING PLAN

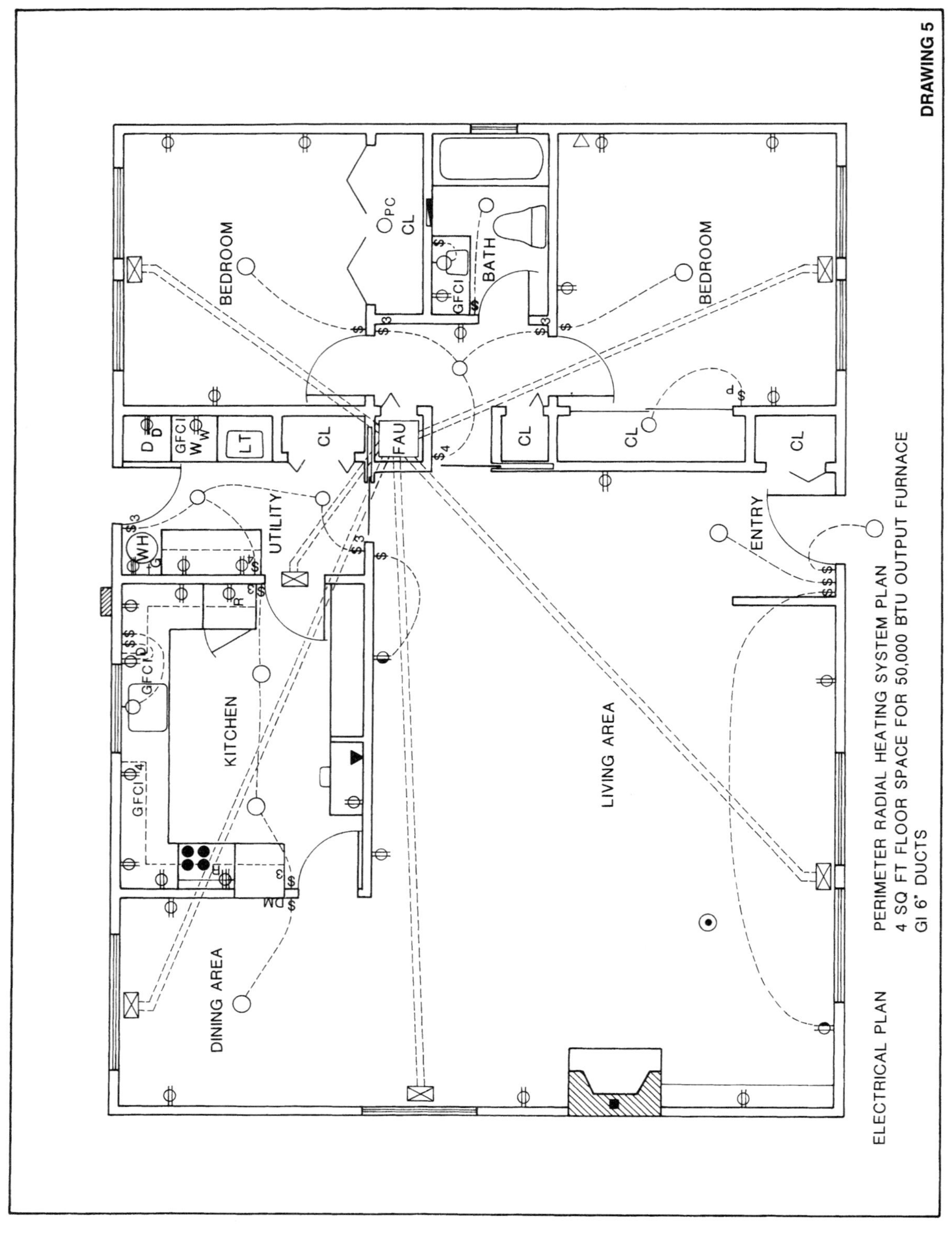

DRAWING 6: FRONT LOADING PLAN AND WINDOW AND DOOR SCHEDULES

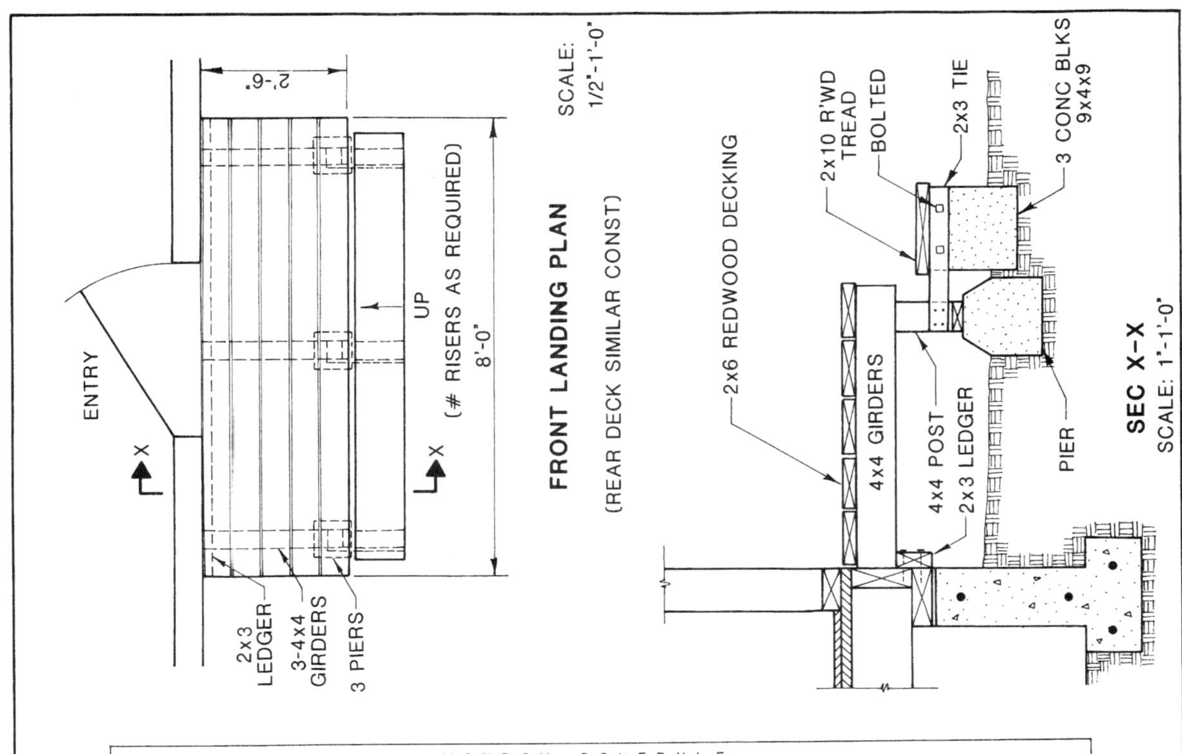

WINDOW SCHEDULE									
SYM	WIDTH	HEIGHT	MATERIAL	TYPE	SCREEN	QUAN	MANUF	CAT #	REMARKS
A	5'-0"	5'-0"	STEEL	AWNING	YES	3	CUSTOM WND MFG	ST-A5	
B	4'=0"	3'-0"	STEEL	SLIDING	YES	4	"	ST-S9	
C	2'-0'	3'-0"	STEEL	LOUVRED	NO	1	"	ST-L7	OBSCURED GLASS
D	6'-0"	5'-0"	STEEL	AWNING	YES	1	"	ST-A9	
E	4'-0"	2'-9"	STEEL	SLIDING	YES	1	"	ST-S7	

DOOR SCHEDULE										
SYM	WIDTH	HEIGHT	THICK.	MATERIAL	TYPE	SCRN	QUAN	MANUF	CAT #	REMARKS
1	3'-0"	6'-8"	1 7/8"	OAK	SOLID CORE	NO	1	CUSTOM DR MFG	EXT-75	4 COATS NAT OIL
2	2'-6"	"	1 7/8"	FIR	SOLID CORE	YES	1	"	EXT-35	2'x3' LOUVERED WND
3	2'-6"	"	1 1/2"	MAHOGANY	HOLLOW CORE	NO	2	"	SLGD-7	SLDG POCKET DOORS
4	2'-6"	"	1 3/8"	MAHOGANY	HOLLOW CORE	NO	4	"	INT-9	2 COATS VARNISH
5	2'-2"	"	1 3/8"	MAHOGANY	HOLLOW CORE	NO	1	"	INT-3	4 COATS VARNISH
6	3'-0"	"	1 3/8"	MAHOGANY	HOLLOW CORE	NO	2	"	CL-80	2 COATS VARNISH
7	21"	"	3/4"	PINE	LOUVERED	NO	4	"	LV-22	PAINT TRIM COLOR
8	7"	"	3/4"	PINE	LOUVERED	NO	4	"	LV-10	PAINT TRIM COLOR
9	12"	"	3/4"	PINE	LOUVERED	NO	4	"	LV-14	PAINT TRIM COLOR
10	6"	"	3/4"	PINE	LOUVERED	NO	4	"	LV-09	PAINT TRIM COLOR

DRAWING 6

DRAWING 7: FRAMING AND INTERIOR ELEVATIONS

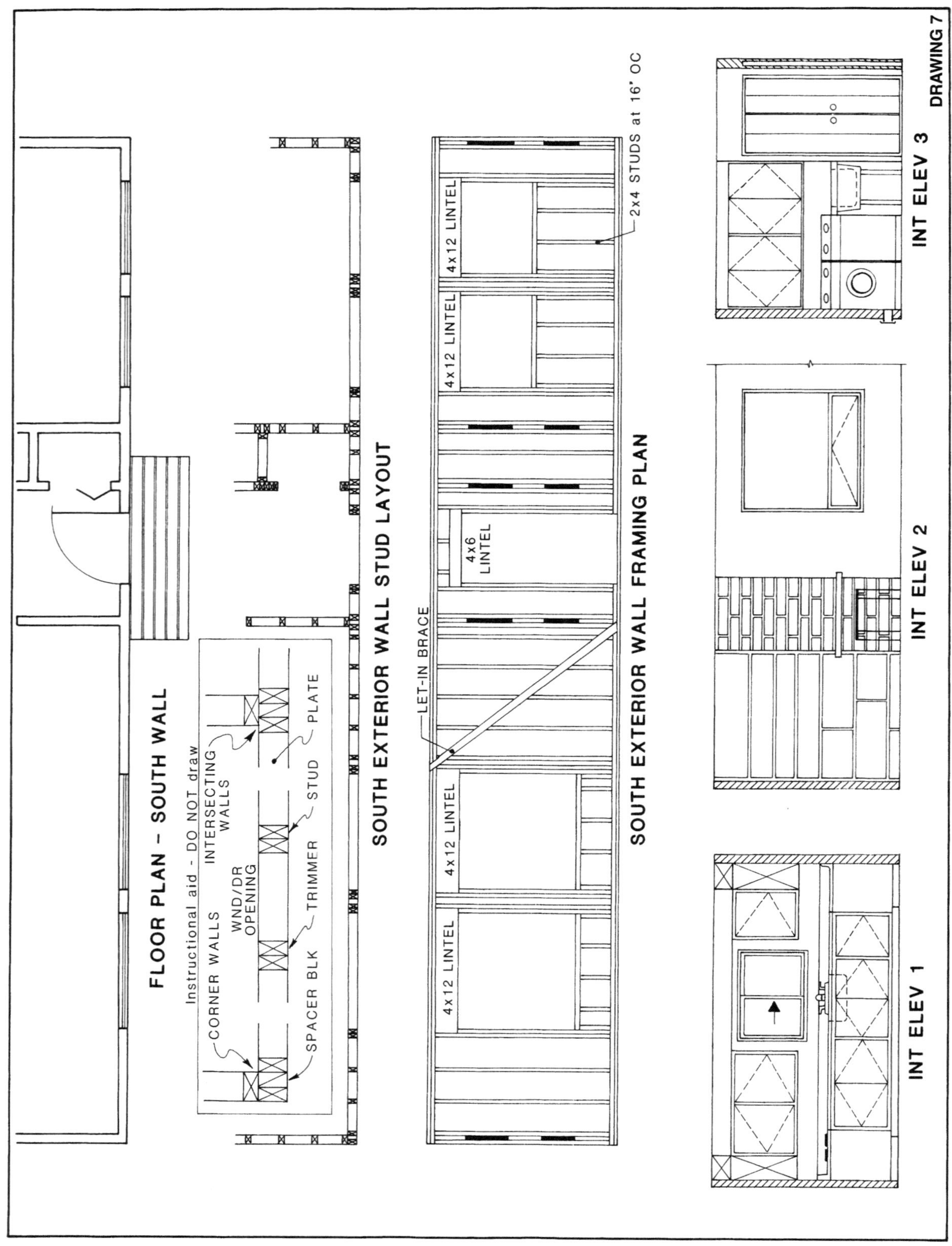

DRAWING 8: SLAB FOUNDATION PLAN

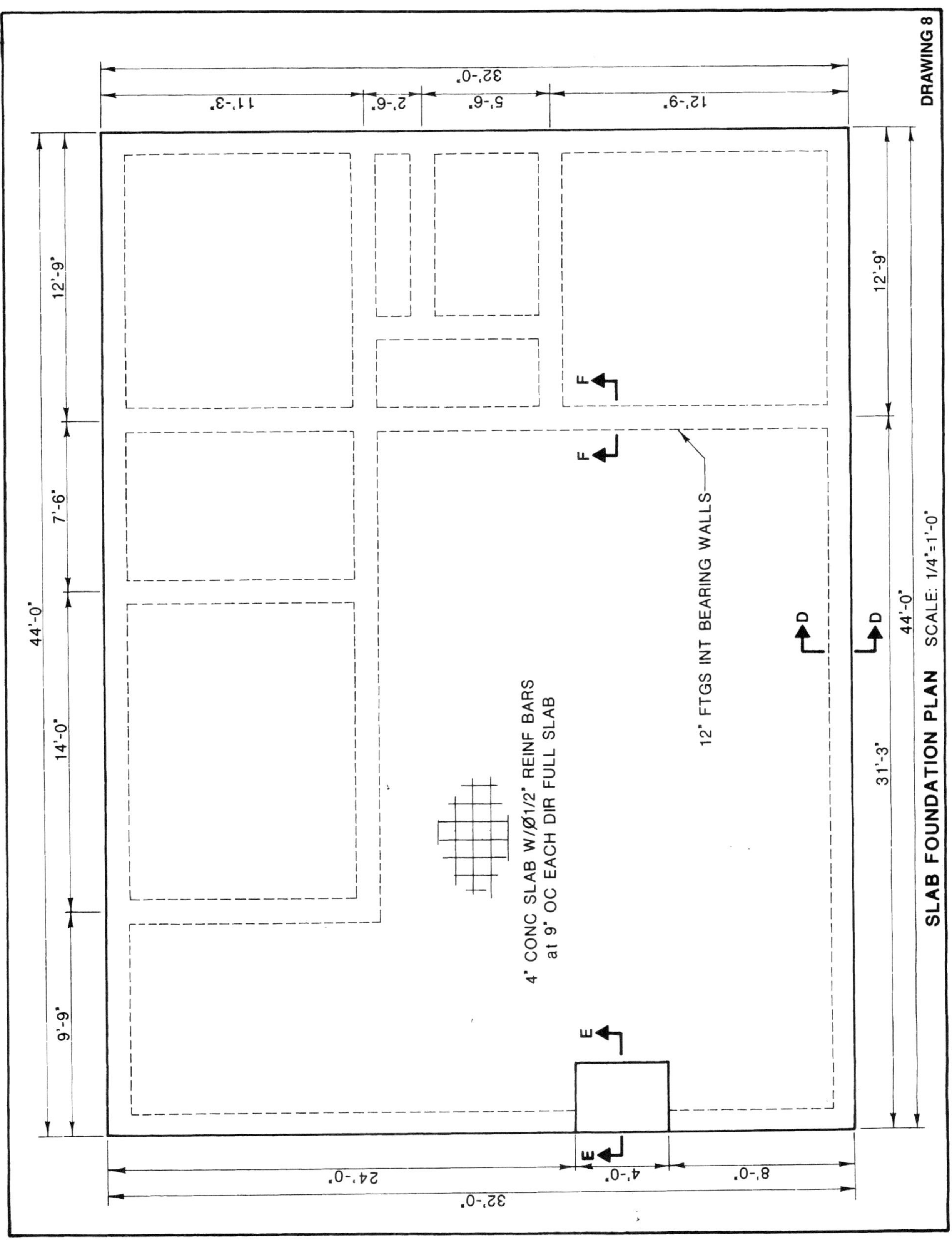

DRAWING 9: T-FOUNDATION PLAN

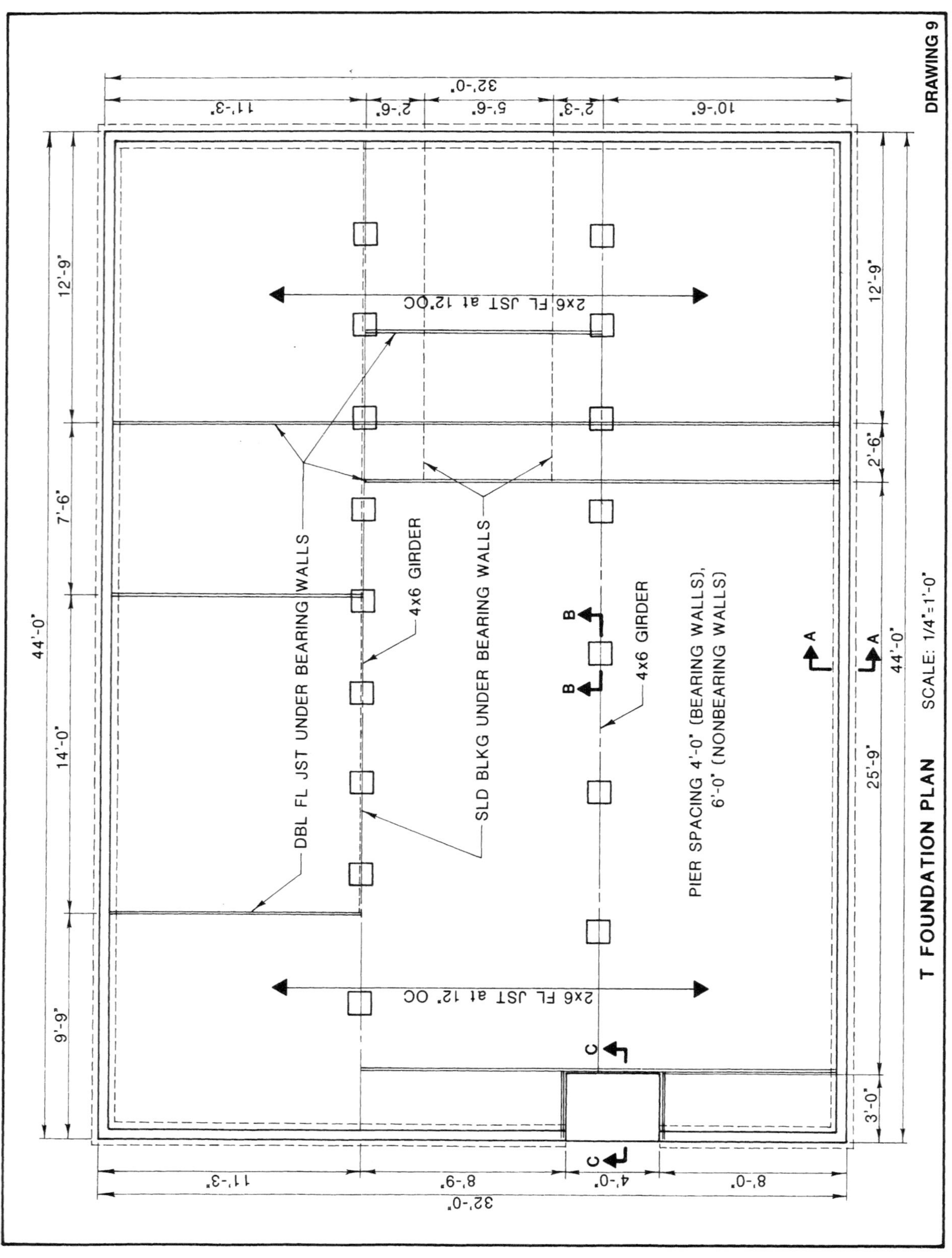

DRAWING 10: CONSTRUCTION DETAILS

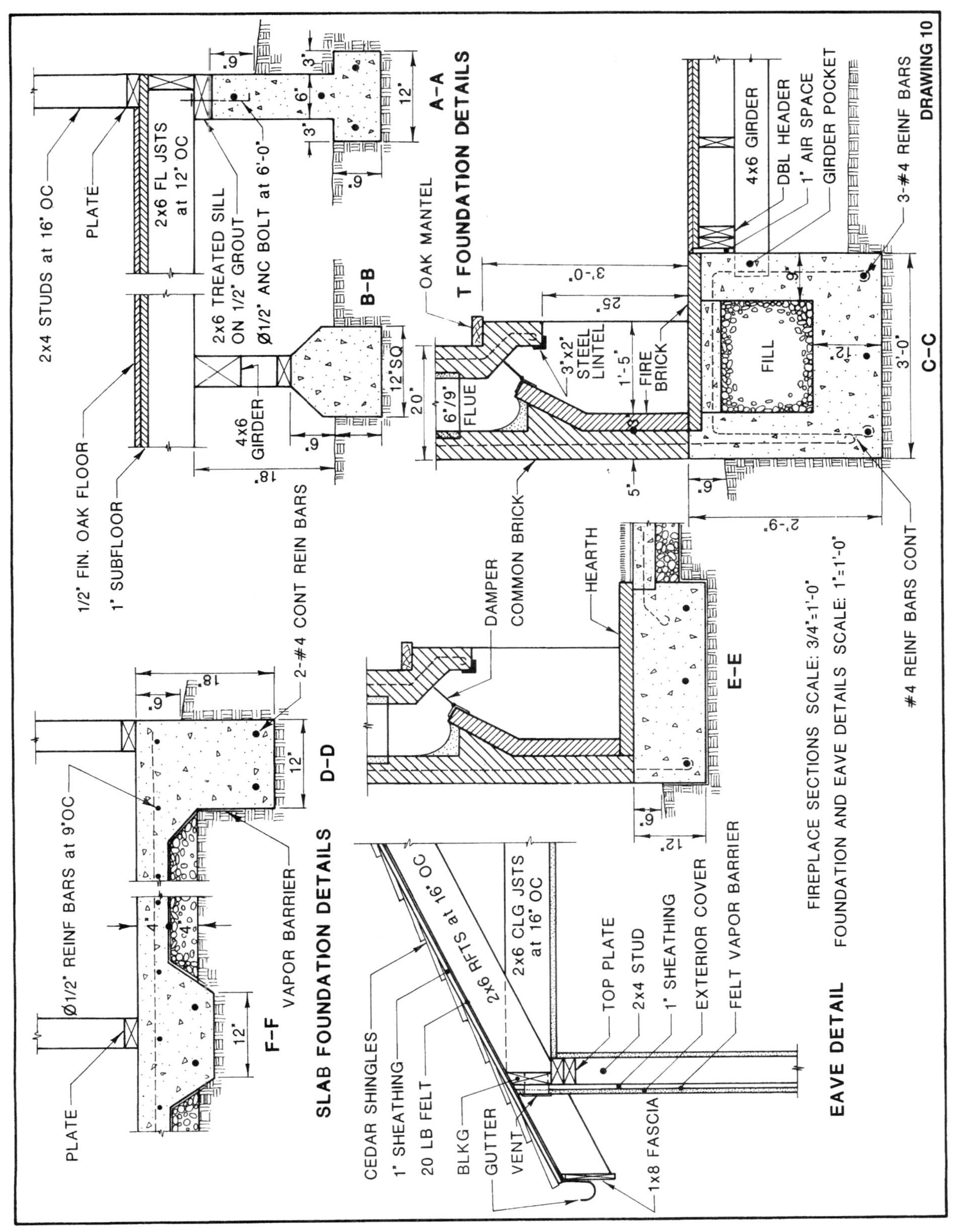

DRAWING 11: GARAGE AND ROOF FRAMING PLANS

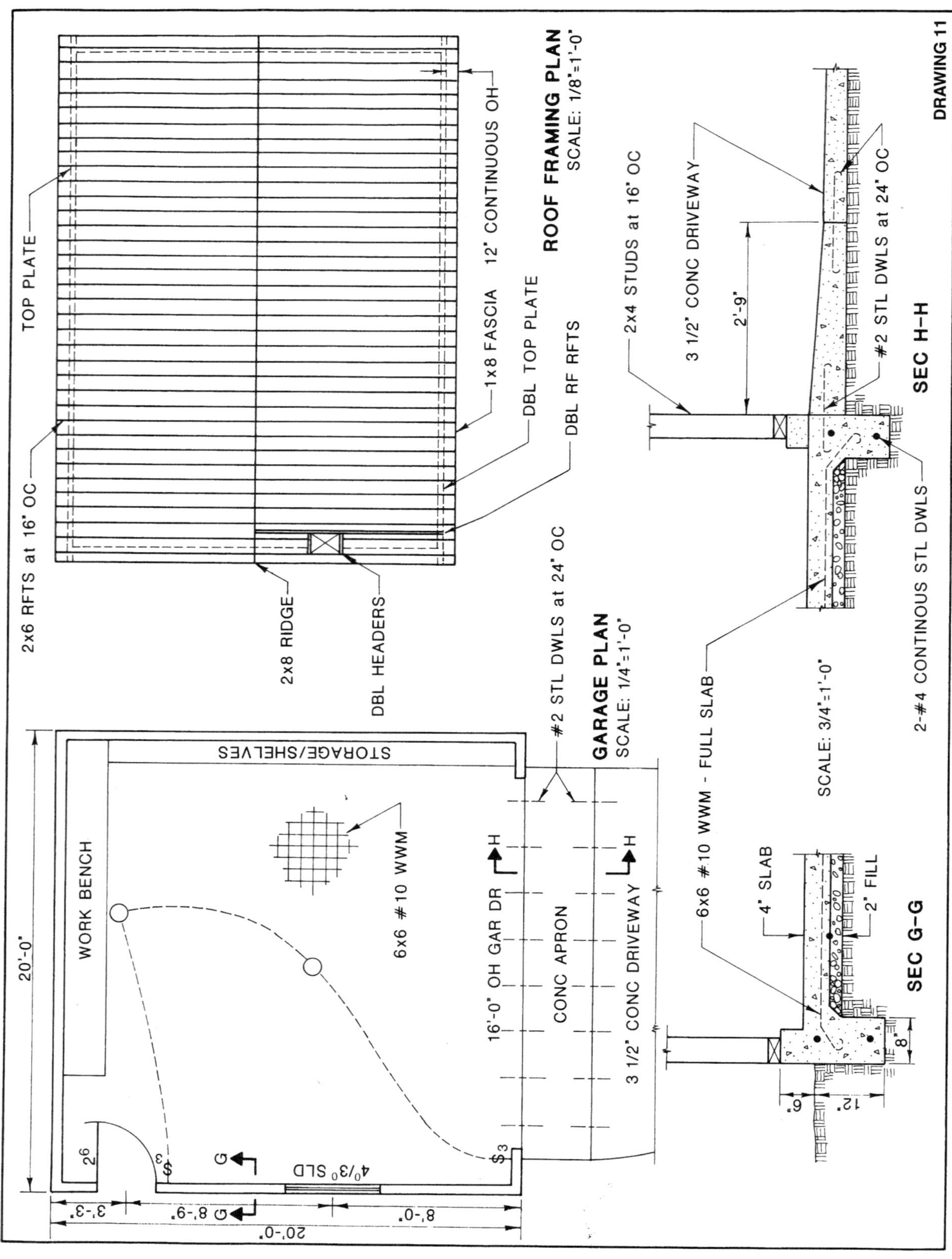

INTERPRETATION TEST ANSWER KEY

Drawing 1: Floor Plan

1. 2″ × 6″
2. 16″
3. on center
4. 3′-0″
5. laminated beam
6. bifold door
7. forced-air unit
8. 
9. hidden lines (dotted)
10. water heater
11. sliding pocket door
12. dryer
13. washer
14. laundry tray (sink)
15. refrigerator
16. flue
17. masonry
18. 44′ × 32′
19. metal joist hangers
20. 

21. 1408 sq.ft.
22. 1/4″ = 1′-0″
23. square
24. circle
25. beams

Drawing 2: Plot Plan

26. 3/32″ = 1′-0″
27. redwood
28. south
29. 20′ × 20′
30. 20′
31. 43′
32. 6′
33. 72′ × 95′
34. 4′
35. roof outline

Drawing 3: East and West Exterior Elevations and Roof Plan

36. continuous overhang
37. 3/32″ = 1′-0″
38. gage galvanized iron
39. fascia board
40. slope diagram
41. 1/4
42. downward slope
43. gable roof
44. 1/4″ = 1′-0″
45. outside edge of exterior walls
46. chimney
47. common brick
48. awning
49. louvered
50. ceiling line

Drawing 4: North and South Exterior Elevations

51. 6′-9″
52. 6′-9″
53. 8′-0″
54. 24″
55. how they are hinged
56. slide direction
57. north elevation
58. 1/4″ = 1′-0″
59. right side
60. slab

Drawing 5: Electrical and Perimeter Radial Heating Plan

61. 
62.
63. $
64.
65.
66.
67.
68. ground fault circuit interrupter
69. perimeter radial forced air
70. $ 3
71.
72. $ DM
73.
74. WH
75. pull cord and switch with pilot light
76. to control a lamp
77. R
78. 6″
79. galvanized iron
80. when the light source is in another room

Drawing 6: Front Landing Plan and Window and Door Schedules

81. 3 concrete blocks
82. wood tie (2″ × 3″)
83. 2″ × 6″
84. 2″ × 10″
85. 2′-6″ × 8′-0″
86. ledger (2″ × 3″)
87. piers or posts
88. girders (4″ × 4″)
89. ST-L7
90. EXT-35

91. obscured glass
92. 6′ × 5′
93. exterior walls
94. 1 7/8″
95. 4 coats of oil

Drawing 7: Framing and Interior Elevations

96. 2″ × 4″
97. 16″
98. 4″ × 12″
99. 4″ × 6″
100. let-in brace
101. 3
102. 3
103. book shelves
104. awning
105. sliding
106. dryer vent
107. cripple studs
108. cripple studs
109. spacer blocks (2″ × 4″)
110. sliding door

Drawing 8: Slab Foundation Plan

111. yes
112. exterior footing
113. fireplace
114. interior footing
115. 12″
116. 4″
117. 1/2″
118. 9″
119. stop cracking
120. 1/4″ = 1′-0″

Drawing 9: T-Foundation Plan

121. 2″ × 6″
122. 12″
123. 6′
124. 4′
125. T-foundation
126. pier
127. fireplace
128. strengthen floor system under bearing walls
129. stops joists from twisting (warping) and provides a nailing surface
130. 4″ × 6″
131. support floor joists
132. 11′-3″ or 10′-9″
133. 11′-3″ or 10′-9″
134. double floor joists
135. solid blocking

Drawing 10: Construction Details

136. damper
137. fire brick
138. fire brick
139. plate
140. 6″
141. 18″
142. cedar shingles
143. 1″ × 8″
144. steel lintel (3″ × 2″)
145. #4
146. 1/2″ oak
147. double headers
148. 1/2″
149. treated sill (2″ × 6″)
150. 12″
151. 6″
152. 1/2″
153. #4 rebars
154. flue
155. 25″
156. 6″ × 9″
157. felt vapor barrier
158. two 2 × 4's
159. fascia (1″ × 8″)
160. 16″
161. 16″
162. 12″
163. 6′
164. 3″
165. girders (4″ × 6″)
166. 20-lb. felt
167. 1″
168. one-half inch of grout
169. 3′
170. girder pocket
171. 3/4″ = 1′-0″
172. 1″ = 1′-0″
173. 2′-9″ T-foundation, 12″ slab foundation
174. 4″
175. 1″ sheathing

Drawing 11: Garage and Roof Framing Plans

176. 2″ × 6″
177. 16″ (on center)
178. 2″ × 8″
179. double headers and double rafters
180. 12″
181. exterior walls (top plate)
182. 2′-9″
183. 3 1/2″
184. welded wire mesh
185. 16′-0″
186. 8″
187. 4″
188. 2″
189. 400 sq. ft.
190. #2
191. 24″

192. 1/8″ = 1′-0″
193. 3/4″ = 1′-0″
194. 1/4″ = 1′ 0″
195. 4′ × 3′
196. 2′-6″
197. #4
198. 2
199. H-H
200. 2 1/2″